Resource Management in Future Internet

RIVER PUBLISHERS SERIES IN COMMUNICATIONS

Volume 38

This series focuses on communications science and technology. This includes the theory and use of systems involving all terminals, computers, and information processors; wired and wireless networks; and network layouts, procontentsols, architectures, and implementations.

Furthermore, developments toward newmarket demands in systems, products, and technologies such as personal communications services, multimedia systems, enterprise networks, and optical communications systems.

- Wireless Communications
- Networks
- Security
- Antennas & Propagation
- Microwaves
- Software Defined Radio

For a list of other books in this series, visit www.riverpublishers.com
http://riverpublishers.com/river publisher/series.php?msg=Communications

Resource Management in Future Internet

Editors

Vladimir Poulkov

Professor, Faculty of Telecommunications
Technical University of Sofia,
Bulgaria

Ramjee Prasad

Director and Professor,
Center for TeleInFrastruktur (CTIF)
Aalborg University,
Denmark

River Publishers
Aalborg

Published, sold and distributed by:
River Publishers
Niels Jernes Vej 10
9220 Aalborg Ø
Denmark

ISBN: 978-87-93102-44-6 (Print)
978-87-93102-45-3 (E book)

Contents

Preface

It is a pleasure to introduce this book "Resource Management in Future Internet" being published by River Publishers. In this book the reader will find collection of chapters from a number of experts working in the field of Future Internet, Internet of Things, Multimedia Networks, Software Defined Radio, Applications and Services. The book intends to provide highlights on the current research and different aspects related to Quality of Service (QoS) and management of resources in multimedia and internet based future networks.

Internet of Things (IoT) is a term widely used to describe uniquely a network of identifiable physical objects (things) and their virtual representations in an Internet-like structure. These objects contain embedded Information & Communication Technology (ICT) to interact or access external environments. Estimates for IoT are huge, since by definition the IoT will be a diffuse layer of devices, sensors, and computing power that will overlay the entire industry. It is expected that the IoT will account for an increasingly number of connections reaching up to 9 billion by 2018. In Chapter 1, the IoT background and technical approaches from different European initiatives are presented, as well as some major activities at project and regulation level related to the roll out of IoT applications.

Besides the IoT the other most prominent ICT paradigm nowadays is cloud computing. Depending on the type of computing resources delivered via the cloud, cloud services take different forms such as Network as a Service, Infrastructure as a service, Platform as a service, Software as a service, Storage as a service, etc. These services are obliged to deliver increased reliability, security, availability and improved QoS. In Chapter 2 the importance of converging the IoT and cloud computing areas is explained. Recommendations to overcome the challenges in current cloud resource management to support big data together with the need for the decentralization of the related mechanisms are given.

In Chapter 3 the capabilities for open access to resource management functions in all IP-based multimedia networks is investigated. The logical architecture for deployment of open access to resource management functions is presented and the standardized capabilities evaluated. In this chapter an approach to the design of Application Programming Interfaces is proposed. Models of QoS resources intended for multimedia traffic are suggested representing the application and the network views on QoS resources. Use cases that illustrate the capabilities for open access to resource management functions in all IP-based multimedia are described.

In future IP multimedia networks QoS provisioning can be considered as a complex task of high importance with many open research issues and design challenges. Chapter 4 is devoted to the call-level performance evaluation in next generation networks, in which dynamic spectrum access and adaptive modulation and coding are implemented for the goal of efficient utilization of transmission resources and high QoS provisioning. The application of cognitive radio and dynamic spectrum access approaches related to QoS provisioning are discussed. The cognitive radio concept is reviewed and proposed as solution for increasing the efficiency in spectrum access and resource allocation. Management functions essential to the operation of the cognitive radio networks for dynamic spectrum allocation are described and explained.

Important aspect in resource management for the provisioning of the required QoS of the users in future networks is related to the possibility of accurate estimation of the position of the users. This is of major importance in next generation mobile networks where power and bandwidth are the resources directly related to ensuring the required QoS, as in most cases the provisioning of these resources is user location dependent in respect to the serving base station. Chapter 5 presents methods and solutions that make possible to overcome the drawbacks of GPS by utilizing 4G technologies for Localization and Positioning. A survey of the available wireless networks technologies is given and a novel method for performing positioning services on mobile devices based on coexistence of WiFi, WiMAX, LTE technologies is proposed.

Chapter 6 is devoted to the diverse set of technologies that are relevant for a Software Communications Architecture. The latter is a specification standard and a component-based software framework for software defined radio. The goal of such architecture is to provide rules and behaviors which make seamless integration of those components into a complete solution. Using component based development allows engineers to concentrate on the application and thus achieve faster time to market and better software reuse.

The last chapter in this book is focused on the challenge of QoS provision for the plethora of Internet services and applications. Key techniques, such as Access Node Control Protocol and Multiprotocol Label Switching are described. The major focus in the chapter is on the key aspects of the transition from Internet-based Multimedia to Multimedia-based Internet. An idea for new flexible transition architecture is presented together with the practical problems that are to be solved.

We would like to thank all the authors for their contributions. Without them this book could not be possible. We hope the readers will enjoy this book and it will benefit them providing additional knowledge and viewpoints in the field of Future Internet.

Editors
R. Prasad
V. Poulkov

List of Figures

List of Tables

1

The Internet of Things

Sofoklis Kyriazakos
Associate Professor, Aalborg University, Denmark
E-mail: Sofoklis Kyriazakos <sk@es.aau.dk>

1.1 Introduction

Kevin Ashton was the first who used the term Internet of Things (IoT) in 1999; "It's not just a "bar code on steroids" or a way to speed up toll roads, and we must never allow our vision to shrink to that scale. The Internet of Things has the potential to change the world, just as the Internet did. May be even more so."[1].

Nowadays, Internet of Things is a term widely used both at academic and business level and describes uniquely identifiable objects (things) and their virtual representations in an Internet-like structure. Estimates suggest that in 5 to 10 years there will be 100 billion devices connected to the Internet. Two orders of magnitude greater than the 1.5 billion PCs and the billion mobile phones that can be connected to the Internet that are today present in the world. [2]

It is expected that the simplest objects soon will dominate the scene. By the end of 2012, for example, physical sensors will generate 20% of non-video Internet traffic.

According to [2] Internet of Things, is a new revolution of the Internet. Objects make themselves recognizable and they get intelligence thanks to the fact that they can communicate information about themselves and they can access information that has been aggregated by other things. All objects can get an active role thanks to their connection to the Internet.

Radio-Frequency Identification (RFID) used to be a prerequisite for the Internet of Things. If all objects and people in daily life were equipped with radio tags, they could be identified and inventoried by computers [3]. Today,

Vladimir Poulkov and Ramjee Prasad (Eds.), Resource Management in Future Internet, 1–20.

RFID is certainly not the only technology that enables an Internet of Things environment.

At a regulation level, the European Commission has issued an action plan [4] describing the Governance of IoT and how to lift obstacles, by adopting a proactive approach.

In the following sections technical characteristics, applications, standardization and policy-making activities in Europe are presented.

1.2 Technical Approach

1.2.1 Addressability and Intelligence

Center was one of the first applications of Internet of Things. The original idea was based on RFID-tags and unique identification through the Electronic Product Code [20].

An alternative view, from the world of the Semantic Web [5] focuses instead on making all things (not just those electronic, smart, or RFID-enabled) addressable by the existing naming protocols, such as URI. The objects themselves do not converse, but they may now be referred to by other agents, such as powerful centralized servers acting for their human owners.

The next generation of Internet applications using Internet Protocol Version 6 (IPv6) would be able to communicate with devices attached to virtually all human-made objects because of the extremely large address space of the IPv6 protocol. This system would therefore be able to identify any kind of object [6].

A combination of these ideas can be found in the current GS1/EPC global EPC Information Services (EPCIS) specifications [7]. This system is being used to identify objects in industries ranging from Aerospace to Fast Moving Consumer Products and Transportation Logistics [8].

Ambient intelligence and autonomous control are not part of the original concept of the Internet of Things. Ambient intelligence and autonomous control do not necessarily require Internet structures, either. However, there is a shift in research to integrate the concepts of the Internet of Things and autonomous control [9]. In the future the Internet of Things may be a non-deterministic and open network in which auto-organized or intelligent entities (Web services, SOA components), virtual objects (avatars) will be interoperable and able to act independently (pursuing their own objectives or shared ones) depending on the context, circumstances or environments.

Embedded intelligence [10] presents an "AI-oriented" perspective of IoT, which can be more clearly defined as: leveraging the capacity to collect and analyze the digital traces left by people when interacting with widely deployed smart things to discover the knowledge about human life, environment interaction, as well as social connection/behavior.

1.2.2 Sizing

The Internet of objects would encode 50 to 100 trillion objects, and be able to follow the movement of those objects. Human beings in surveyed urban environments are each surrounded by 1000 to 5000 trackable objects [11].

In this Internet of Things, made of billions of parallel and simultaneous events, time will no more be used as a common and linear dimension [12] but will depend on each entity (object, process, information system, etc.). This Internet of Things will be accordingly based on massive parallel IT systems (Parallel computing).

In an Internet of Things, the precise geographic location of a thing—and also the precise geographic dimensions of a thing—will be critical [13]. Currently, the Internet has been primarily used to manage information processed by people. Therefore, facts about a thing, such as its location in time and space, have been less critical to track because the person processing the information can decide whether or not that information was important to the action being taken, and if so, add the missing information (or decide to not take the action). (Note that some things in the Internet of Things will be sensors, and sensor location is usually important. [14]) The GeoWeb and Digital Earth are promising applications that become possible when things can become organized and connected by location. However, challenges that remain include the constraints of variable spatial scales, the need to handle massive amounts of data, and an indexing for fast search and neighbour operations. If in the Internet of Things, things are able to take actions on their own initiative, this human-centric mediation role is eliminated, and the time-space context that we as humans take for granted must be given a central role in this information ecosystem. Just as standards play a key role in the Internet and the Web, geospatial standards will play a key role in the Internet of Things.

1.2.3 Architecture

In the following sections we present architectural proposals and approaches, as well as Internet of Things project solutions.

1.2.3.1 IOT-A

IoT-A, is a European Project that addresses the Internet-of-Things, see Figure 1.1, Architecture. IoT-A proposes the creation of an architectural reference model together with the definition of an initial set of key building blocks. Together they are envisioned as crucial foundations for fostering a future Internet of Things. Using an experimental paradigm, IoT-A combines top-down reasoning about architectural principles and design guidelines with simulation and prototyping to explore the technical consequences of architectural design choices [15, 16].

The IoT reference architecture, see Figure 1.2, is the reference for building compliant IoT architectures and it provides views and perspectives on different architectural aspects that are of concern to stakeholders of the IoT. The creation of the IoT Reference architecture focuses on abstract sets of mechanisms rather than concrete application architectures.

In [17] IOT-A presents the methodology and the initial architectural reference model for Internet of Things. In the following figure, one can see the seven functional groups defines, namely:

- **Applications** that are built on top of an implementation of the IoT-A architecture.
- **Process execution and service orchestration** so that IoT Services become available to external entities and can be composed by them.
- **Virtual entity (VE) and information** enabling search for services.

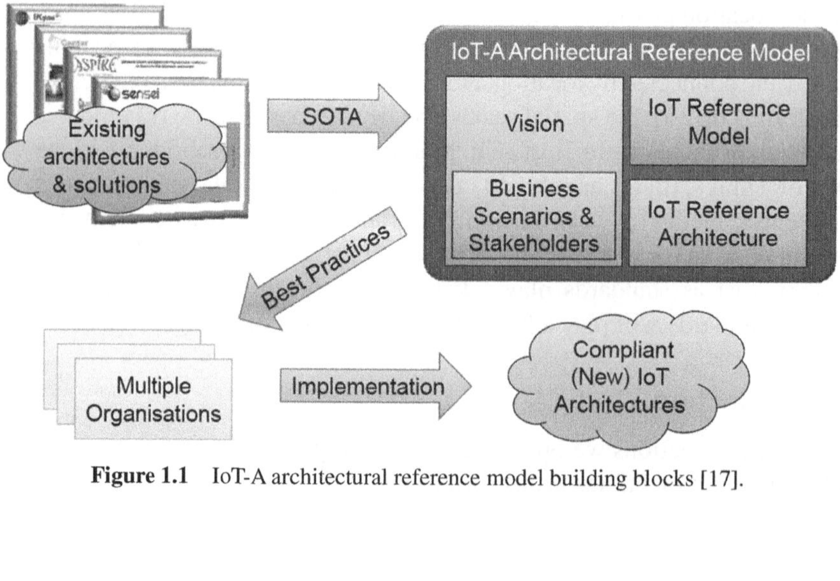

Figure 1.1 IoT-A architectural reference model building blocks [17].

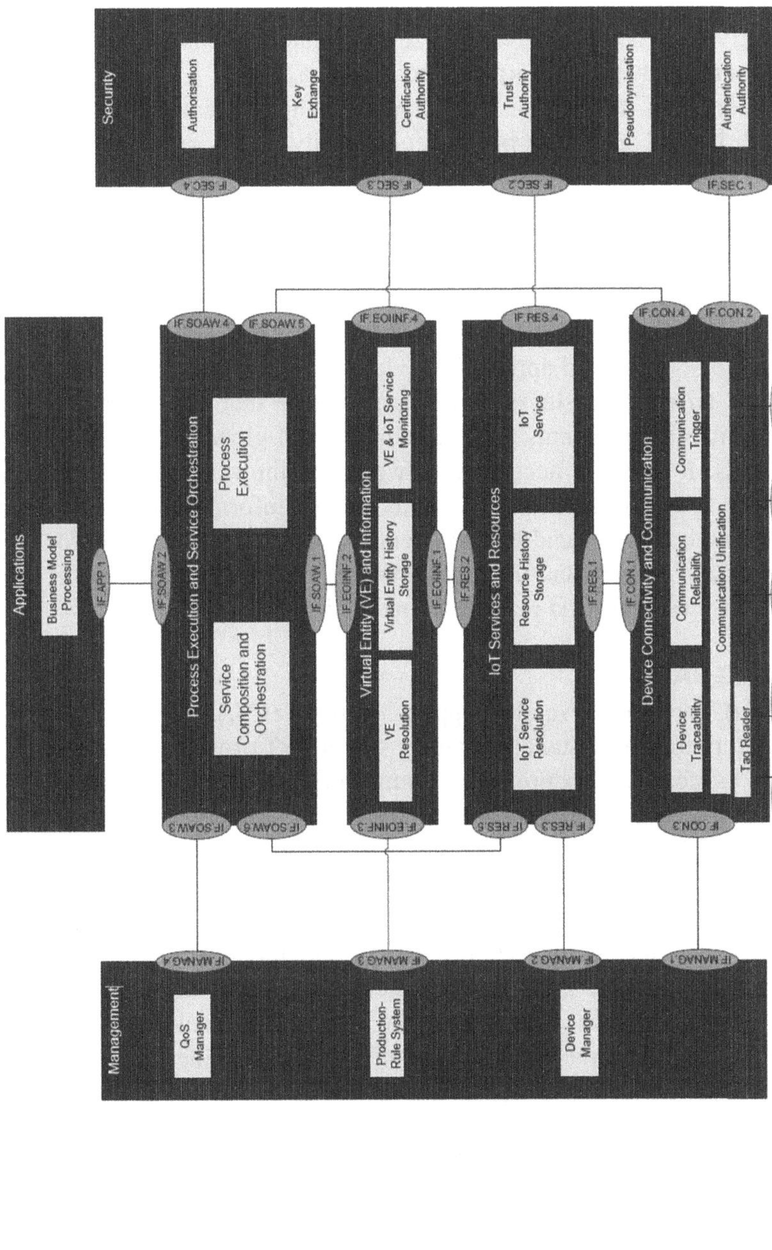

Figure 1.2 functional view of the IoT reference architecture [17].

- **IoT service & resource** providing links to the exposed resources and functionalities required by services for processing information and for notifications.
- **Device connectivity and communication** providing the set of methods and primitives for device connectivity and communication.
- **Management** by a single group of functionalities.
- **Security** functions to be consistently applied by the different groups of functionalities and privacy through pseudonymity.

1.2.3.2 SENSAI

In order to realise the vision of Ambient Intelligence in a future network and service environment, heterogeneous wireless sensor and actuator networks (WS&AN) have to be integrated into a common framework of global scale and made available to services and applications via universal service interfaces. SENSEI creates an open, business driven architecture that fundamentally addresses the scalability problems for a large number of globally distributed WS&AN devices. It provides necessary network and information management services to enable reliable and accurate context information retrieval and interaction with the physical environment. By adding mechanisms for accounting, security, privacy and trust it enables an open and secure market space for context-awareness and real world interaction [21].

1.2.3.3 CASAGRAS

CASAGRAS (Coordination and Support Action (CSA) for Global RFID-related Activities and Standardisation) was a FP7 project that ended in June 2009. Its goal was to provide a framework of foundation studies to assist the European Commission and the global community in defining and accommodating international issues and developments concerning radio frequency identification (RFID) with particular reference to the emerging Internet of Things.

CASAGRAS2 started in June 2010 and will end June 2012. The consortium consists of partners from Europe, the USA, China, Japan, Brazil and Korea. The stated goal is: "To address the key international issues that are important in providing the foundations and co-operation necessary for realising the Internet of Things as a global initiative" [22].

As a CSA, the project did not provide available technological output. Instead CASAGRAS project aim to collect, review and analyse current and emerging proposals and solutions in the IoT domain.

1.2.3.4 Smart santander

Smart Santander proposes a unique in the world city-scale experimental research facility supporting typical applications and services for a smart city. The facility will be open to FIRE community and will provide an infrastructure for IoT experimentation at different levels (communication protocols, IoT architectures, services to end-users, etc.).

Smart Santander will provide access to around 20.000 IoT devices deployed mainly in the city of Santander but also in other places worldwide (Guilford, Belgrade, Australia, Lübeck, Aarhus, etc.) [23].

1.2.3.5 BRIDGE

The EPC Information Services is a role defined in EPC global Network Architecture Framework for storage and retrieval of filtered and processed information about different events within the supply-chain. The EPCIS offers two interfaces: one for query request and the other one for capture operations. The query interface allows trading partners to query information about any event data stored in the EPCIS-repository together with business context. However for such a decentralized architecture, since the complete information about an individual object may be fragmented across multiple organizations, there is a need for secure lookup services for locating all the providers of the fragments of information that constitute the complete supply-chain or lifecycle history for an object [9].

To enable RFID and EPC global standard solutions in practice, technical, social and educational constraints - particularly in the area of security must be overcome. BRIDGE (Building Radio frequency Identification solutions for the Global Environment) addresses these problems by extending the EPC network architecture. This is done by researching, developing and implementing tools which will enable the deployment of EPC global applications in Europe. The “enablement” is mostly in the development of security apparatus, both in hardware, software and business practises.

1.2.3.6 Smart products

Smart Products develops the scientific and technological basis for building “smart products” with embedded —proactive knowledge. Proactive knowledge encompasses knowledge about the product itself (features, functions, dependencies, usage, etc.), its environment (physical context, other smart products) and its users (preferences, abilities, intentions, etc.).

Therefore, smart products —talk||, —guide||, and —assist|| designers, workers and consumers dealing with them. Some proactive knowledge will be co-constructed with the product, while other parts are gathered during the product lifecycle using embedded sensing and communication capabilities. The outcome of Smart Products will impact the manufacturing and consumer domain, primarily targeting consumer goods, automotive and aerospace industries, spanning both product innovation (for consumer goods and automotive) and process innovations (for automotive and aerospace). [24]

1.2.3.7 CUBIQ

The Cross UBiQuituous Platform (CUBIQ) project aims to develop a common platform that facilitates the development of context-aware applications. The idea is to provide an integrated platform that offers unified data access, processing and service federation on top of existing, heterogeneous ubiquitous services. The CUBIQ architecture consists of three layers: (1) a data resource layer, (2) an intra-context processing layer and (3) an inter-context processing layer. The data resource layer provides transparent data access and handles mobility, migration, replication, concurrency, faults and persistency. The intra-context layer provides data processing services. The inter-context processing layer is responsible for service composition. The CUBIQ architecture provides interfaces for each layer.

Technically, CUBIQ uses PIAX, an open source agent platform for P2P structured overlay networks and a Universal Service Description Language (USDL) for describing services in CUBIQ. Further CUBIQ provides solutions for real-time stream data processing and complex event processing.

1.3 Applications

There are several applications, where IoT can significantly improve services and services. In the following figure the major IoT applications have been grouped into 9 categories.

The intense use of devices (Things) in a way that these are addressable, allows developers to exploit feeds and develop applications that increase productivity and efficiency in business processes, allows environmental monitoring and early warnings, assists patients through advanced eHealth applications and allows the development of smart cities and buildings.

Pachube, recently rebranded to Cosm, is one of the pioneers in IoT applications since 2008, being open and sharing feeds from devices registered inits platform. Today, Cosm is the platform, API and community where devices, information, developers, apps and commercial applications come together to bring connected products and ideas to life. Cosm is a secure, scalable platform

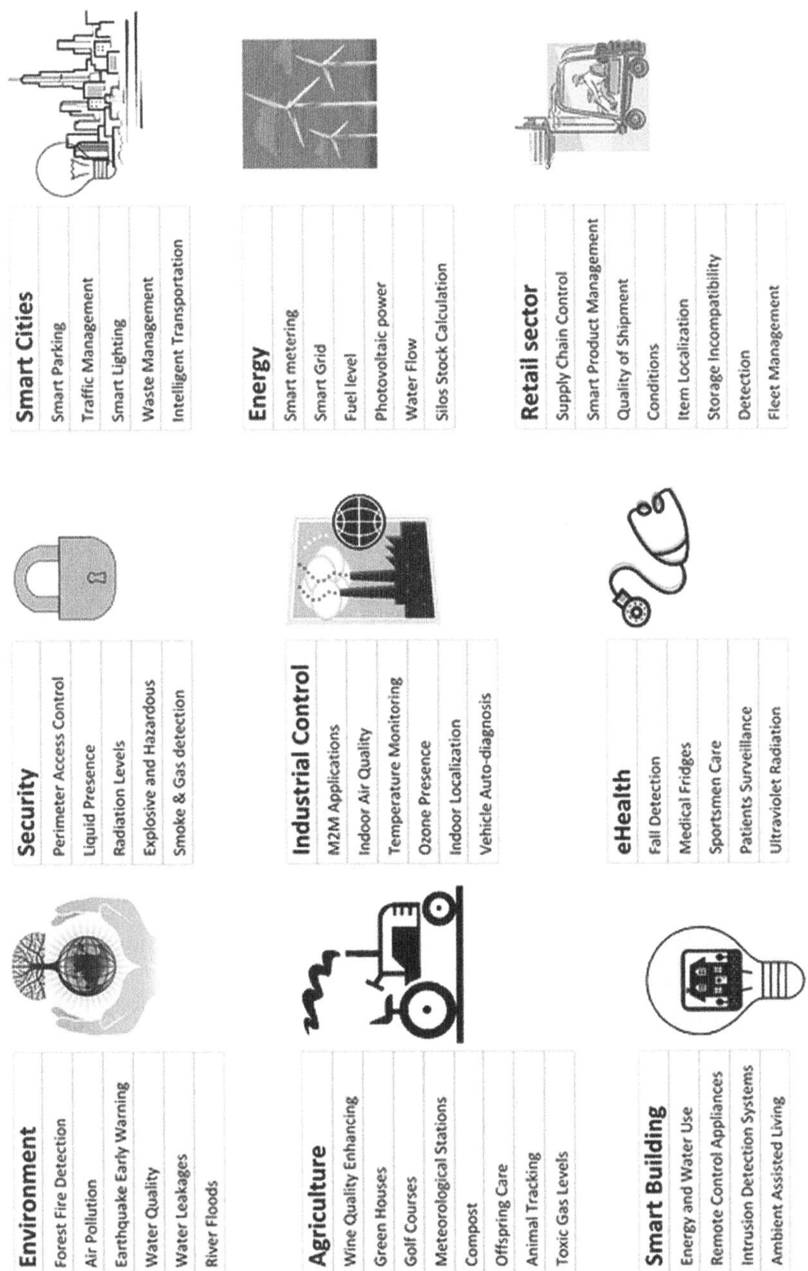

Figure 1.3 IoT applications.

that connects devices and products with applications to provide real-time control and data storage. Using Cosm's open API, individuals and companies can create new devices, develop prototypes, and bring products to market in volume. Cosm offers a way to launch internet enabled products without having to build any backend infrastructure [19].

1.4 Standardization Activities

1.4.1 ETSI TC M2M

The European Telecommunications Standard Institute created in January 2009 a new Technical Committee (TC) focused on Machine-to-Machine communications. The main responsibilities of the group can be summarized in the following:

- to collect and specify M2M requirements from relevant stakeholders;
- to develop and maintain an end-to-end overall high level architecture for M2M;
- to identify gaps where existing standards do not fulfill the requirements and provide specifications and standards to fill these gaps, where existing standards bodies or groups are unable to do so;
- to provide the ETSI main centre of expertise in the area of M2M;
- to co-ordinate ETSI's M2M activity with that of other standardization groups

1.4.2 ITU-T USN

Within ITU-T, USN standardization is being carried out under the auspices of the Next-Generation Network Global Standards Initiative (NGN-GSI).

According to ITU-T, USN is a conceptual network built over existing physical networks which make use of sensed data and provide knowledge services. So its main components are the following ones:

- USN Applications and Services platform: technology platform to enable the effective use of a USN in a given application or service.
- USN Middleware: including functionalities for sensor network management and connectivity, event processing, sensor data mining, etc.
- Network infrastructure: mainly based on Next-Generation-Networks (NGNs), USN is not a physical network it is a conceptual network making use of existing networks.
- USN Gateway: A node which interconnects sensor networks with other networks.

- Sensor network: Network of inter-connected sensor nodes IP based nodes possibility of direct connection to NGN; non IP based nodes, often managed via gateway).

ITU-T work has been focused mainly on general functional models and requirements to support of USN application and services in NGN environments. A reference model of USN over NGN has been proposed based on both USN service requirements, and requirements of NGN (enhanced or additional) necessary to support USN applications and services. ITU-T recommendations also include a general functional model and requirements for USN middleware, as well as management and security issues.

1.4.3 ISO/IEC JTC1 WG7 (Working Group on Sensor Networks)

ISO is the International Organization for Standardization, current JTC (Joint Technical Committee) 1/WG (Work Group) 7 has been preceded by JTC 1 SGSN SC6, the ISO/IEC Study Group on Sensor Networks SGSN that was established in 2007.

A result of JTC 1 SGSN activities was the ISO/IEC 29182 Reference architecture for sensor networks application and services. This work area can give them an overall architecture consisting of system and network configurations, data processing functionalities, and interface relationships due to heterogeneous application components. The ISO/IEC 29182 Reference architecture was transferred to JTC 1 WG7 (established in October 2009).

Main Terms of Reference or study items defined under JTC 1 WG7 are: 1) generic solutions for sensor networks, 2) application oriented sensor networks 3) to foster communication and sharing of information between groups.

The recent creation of ISO/IEC JTC1 WG7 (Working Group on Sensor Networks) makes its activities still preliminary and very general. However it is expected a rapid growth both in future activities and relevance in the IoT standardization field in liaison with other important SDOs (as ITU-T USN).

1.4.4 OGC® SWE (Sensor Web Enablement)

Open Geospatial Consortium, OGC®, overall activity is the so-called Geospatial Web: to publish, discover and use geo data as well as geo processing services in an interoperable way.

OGC® Sensor Web Enablement (SWE) mission is the development of standards allowing the integration of sensors and sensor networks into the Geospatial Web (i.e. it is a specialized subtype called the —Sensor Web).

SWE standardization activity is being pursued through the establishment of several encodings for describing sensors and sensor observations, and through several standard interface definitions for web services.

Main functionalities in OGC® SWE are:

- Quickly discover sensors and sensor data (secure or public) that can meet different needs based on location, observables, quality, ability to task, etc.
- Description of sensors and measurements in a standard machine understandable encoding enabling assessment and processing without a-priori knowledge.
- Readily to access sensor parameter in a common manner, and in a highly configurable way.
- Access to measurement data (real-time data as well as time series) based on standardized data format.
- Tasking of sensors to acquire observations of interest.
- Alerting mechanisms based on sensor measurements and certain alert criteria: subscribe to and receive alerts when a sensor measures a particular phenomenon.

Within the last years, the SWE architecture has been advanced to a solid and mature state. Now, most of the SWE standards have been adopted as official OGC® standards and several practical applications relying on the SWE standards have been built.

1.4.5 IETF (constrained devices)

The Internet Engineering Task Force, IETF, is an international community of network designer, operators, vendors, and researchers concerned with the evolution of the Internet architecture and the smooth operation of the Internet. For efficient organization, the technical work of IETF is divided into various working groups, which are organized into several research areas such as routing, transport, or security. The mailing list represents one of the most important tools for managing work and the IETF meets three times a year.

The IETF working groups are grouped into areas, and managed by Area Directors (ADs), members of the Internet Engineering Steering Group (IESG). The Internet Architecture Board (IAB), that providing architectural oversight, and IESG are chartered by the Internet Society (ISOC) for these purposes. The mission of the IETF is to make the Internet work better by producing high quality, relevant technical documents that influence the way people design, use, and manage the Internet. In order to realize its mission, IETF follows these basic principles:

- Open process;
- Technical competence;
- Volunteer Core;
- Rough consensus and running code;
- Drafts ownership.

The basic definition of the IETF standards process is defined in to RFC202.

1.4.6 EPC Global Network Architecture

In recent years, the Electronic Product Code (EPC) — a worldwide, unambiguous code for the designation of physical goods — has become the subject of enormous interest, not only in research but also in several industries and society in general. The rapid and escalating diffusion of the EPC was particularly driven by the Auto-ID Center, a project to develop RFID standards founded in 1999 at the Massachusetts Institute of Technology (MIT) with cooperation from numerous industrial sponsors. The Auto-ID Center created the EPC to ensure RFID interoperability in supply-chain-wide applications. An important feature is its capability to serve as a meta scheme that integrates with existing numbering schemes, such as the serialized Global Trade Item Number (GTIN) standard used in retail. However, the Center‘s long-term objective wasn‘t only standardizing numbering formats but also developing an entire family of open standards, including air interface protocols, software interfaces, and directory services, to bridge the gap between the physical and virtual worlds. In October 2003, the Auto-ID Center was transformed into an international research network known as Auto-ID Labs, which concentrates on technology as well as application-oriented research, and EPC global, a nonprofit organization responsible for commercializing, standardizing, and managing EPC standards.

1.4.7 REST Architecture

REST is a coordinated set of architectural constraints that attempts to minimize latency and network communication, while at the same time maximizing the independence and scalability of component implementations. This is achieved by placing constraints on connector semantics, where other styles have focused on component semantics. REST enables the caching and reuse of interactions, dynamic substitutability of components, and processing of actions by intermediaries, in order to meet the needs of an Internet-scale distributed hypermedia system.

REST elaborates only those portions of the architecture that are considered essential for Internet-scale distributed hypermedia interaction. Areas for improvement of the Web architecture can be seen where existing protocols fail to express all of the potential semantics for component interaction, and where the details of syntax can be replaced with more efficient forms without changing the architecture capabilities. Likewise, proposed extensions can be compared to REST to see if they fit within the architecture; if not, it is usually more efficient to redirect that functionality to a system running in parallel with a more applicable architectural style.

1.4.8 Web Service Architecture

The WSA provides a conceptual model and a context for understanding Web services and the relationships between the components of this model. The architecture does not attempt to specify how Web services are implemented, and imposes no restriction on how Web services might be combined. The WSA describes both the minimal characteristics that are common to all Web services, and a number of characteristics that are needed by many, but not all, Web services. The Web services architecture is interoperability architecture: it identifies those global elements of the global Web services network that are required in order to ensure interoperability between Web services.

Web services are most appropriate for applications:

- That must operate over the Internet where reliability and speed cannot be guaranteed;
- Where there is no ability to manage deployment so that all requesters and providers are upgraded at once;
- Where components of the distributed system run on different platforms and vendor products;
- Where an existing application.

1.5 Policy-making in Europe

Many of these changes that will be required in the adoption of IoT in the EU member states will have to be addressed by European policy-makers and public authorities to ensure that the use of IoT technologies and applications will stimulate economic growth, improve individuals' well-being and address some of today's societal problems.

Therefore, EC has prepared for the EU Parliament An action plan for Europe where the following line of actions are presented [7]:

Line of action 1 — Governance

The Commission will initiate and promote, in all relevant fora, discussions and decisions on

- defining a set of principles underlying the governance of IoT;
- setting up an 'architecture' with a sufficient level of decentralized management, so that public authorities throughout the world can exercise their responsibilities as regards transparency, competition and accountability.

Line of action 2 — Continuous monitoring of the privacy and the protection of personal data questions.

The Commission recently adopted a Recommendation 29 that provides guidelines on how to operate RFID applications in compliance with privacy and data protection principles

These two examples illustrate how, in practice, the Commission will watch over the application of data protection legislation to IoT:

- by consulting, when necessary, the Article 29 Data Protection Working Party;
- by providing guidance on the correct interpretation of EU legislation;
- by fostering dialogue among stakeholders;
- by proposing, if necessary, additional regulatory instruments.

Line of action 3 — The 'silence of the chips'

The Commission will launch a debate on the technical and legal aspects of the 'right to silence of the chips', which has been referred to under different names by different authors and expresses the idea that individuals should be able to disconnect from their networked environment at any time.

Line of action 4 — Identification of emerging risks

The Commission will follow the ENISA work mentioned above and will take further action as appropriate, including regulatory and non-regulatory measures, to provide a policy framework that enables IoT to meet the challenges related to trust, acceptance and security.

Line of action 5 — IoT as a vital resource to economy and society

Should IoT grow to the importance it is expected to attain, any disruption might have a significant impact on economy and society. The Commission will therefore closely follow the development of IoT infrastructures into a vital resource for Europe, especially in connection with its activities on the protection of critical information infrastructure 33.

Line of action 6 — Standards Mandate

The Commission will assess the extent to which existing standards mandates can include further issues related to IoT35 or launch additional mandates if necessary. Additionally, the Commission will keep monitoring developments in European Standards Organisations (ETSI, CEN, CENELEC), their international counterparts (ISO, ITU) and other standards bodies and consortia (IETF, EPC global, etc) with a view for IoT standards to be developed in an open, transparent and consensual manner with the participation of all interested parties.

Line of action 7 — Research and Development

The Commission will continue to finance FP7 research projects in the area of IoT, putting an emphasis on important technological aspects such as microelectronics, non-silicon based components, energy harvesting technologies, ubiquitous positioning, networks of wirelessly communicating smart systems, semantics, privacy- and security-by-design, software emulating human reasoning and on novel applications.

Line of action 8 — Public-Private Partnership

The Commission is currently preparing the setting-up of four public-private partnerships (PPP) where IoT can play an important role. Three of them, 'green cars', 'energy-efficient buildings' and 'Factories of the Future' were proposed by the Commission as part of the recovery package 38. The fourth one, 'Future Internet', aims at further integrating the existing ICT research efforts in relation to the future of the Internet39.

Line of action 9 — Innovation and pilot projects

Complementing the research activities listed above, the Commission will consider promoting the deployment of IoT applications by launching pilot projects through CIP41. These pilots should focus on IoT applications that deliver strong benefits to society, such as e-health, e-accessibility, climate change, or helping to bridge the digital divide.

Line of action 10 — Institutional Awareness

The Commission will regularly inform the European Parliament, the Council, the European Economic and Social Committee, the Committee of the Regions, the Article 29 Data Protection Working Party 42 and any other relevant stakeholders about IoT developments.

Line of action 11: International dialogue

The Commission intends to intensify the existing 43, 44 dialogue on all aspects of IoT with its international partners, aiming to agree on relevant joint actions, share best practices and promote the lines of action laid down in this Communication.

Line of action 12 — RFID in recycling lines

As part of its regular monitoring of the waste management industry, the Commission will launch a study to assess the difficulties of recycling tags and the benefits and nuisances that the presence of tags can have on the recycling of objects.

Line of action 13 — Measuring the uptake

Eurostat will start publishing in December 2009 statistics on the use of RFID technologies. Monitoring the introduction of IoT related technologies will provide information on their degree of penetration and allow the assessment of their impact on the economy and the society as well as the effectiveness of the related Community policies.

Line of action 14 — Assessment of evolution

Beyond the specific aspects mentioned above, it is important that a multi-stakeholder mechanism is put in place at European level to:

- monitor the evolution of IoT;
- support the Commission in carrying out the various actions listed in this Communication;
- assess which additional measures should be undertaken by European Public Authorities.

The Commission will use FP7 to conduct this work, by gathering a representative set of European stakeholders and ensuring a regular dialogue and sharing of best practices with other world regions.

1.6 Conclusions

In the sections of this chapter, the Internet of Things background and technical approaches from different initiatives have been presented. Emphasis is given on European activities both at project and regulation level, as these are important means to achieve a roll-out of IoT applications. These applications, as presented above, cover a large spectrum of activities and several business sectors, thus proving that the potential is big and the benefits for the community even bigger. It is very promising that there is structured work going on at R&D level, in order to converge to architectures and frameworks that will be the basis for a long-term exploitation, aiming to be prepared for the new digital era.

References

[1] Kevin Ashton: That 'Internet of Things' Thing. In: RFID Journal, 22 July 2009. Retrieved 8 April 2011.
[2] Casaleggio Associati The Evolution of Internet of Things 2011.
[3] P. Magrassi, T. Berg, A World of Smart Objects, Gartner research report R-17-2243, 12 August 2002.
[4] Commission of the European Communities (18 June 2009). "Internet of Things — An action plan for Europe" (PDF). COM (2009) 278 final.
[5] Dan Brickley et al., c. 2001.
[6] Waldner, Jean-Baptiste (2008). Nanocomputers and Swarm Intelligence. London: [[ISTE (publisher)]]. pp. p227–p231. ISBN 1-84704-002-0.
[7] http://www.gs1.org/gsmp/kc/epcglobal/epcis.
[8] Miles, Stephen B. (2011). RFID Technology and Applications. London: Cambridge University Press. pp. 6–8. ISBN 978-0-521-16961-5.
[9] Uckelmann, Dieter; Isenberg, Marc-André; Teucke, Michael; Halfar, Harry; Scholz-Reiter, Bernd (2010). "An integrative approach on Autonomous Control and the Internet of Things". In Ranasinghe, Damith; Sheng, Quan; Zeadally, Sherali. Unique Radio Innovation for the 21st Century: Building Scalable and Global RFID Networks. Berlin, Germany: Springer. pp. 163–181. ISBN 978-3-642-03461-9. Retrieved 28 April 2011.
[10] "Living with Internet of Things, The Emergence of Embedded Intelligence (CPSCom-11)". Bin Guo. Retrieved 6 September 2011.
[11] Waldner, Jean-Baptiste (2007). Nanoinformatique et intelligence ambiante. Inventer l'Ordinateur du XXIeme Siècle. London: Hermes Science. pp. p254. ISBN 2-7462-1516-0.
[12] Janusz Bucki, "L'organisation et le temps" (in French).
[13] Open Geospatial Consortium, "OGC Abstract Specification".
[14] Mike Botts et al, "OGC Sensor Web Enablement: Overview And High Level Architecture".
[15] IOT-A www.iot-a.eu.

[16] IOTA-D1.1 "SOTA report on existing integration frameworks/ architectures for WSN, RFID and other emerging IoT related Technologies".
[17] IoTA D-1.2 "Initial Architectural Reference Model for IoT".
[18] IOT-I www.iot-i.eu.
[19] http://cosm.com.
[20] http://en.wikipedia.org/wiki/Internet_of_Things.
[21] http://www.ict-sensei.org/.
[22] http://www.iot-casagras.org/.
[23] http://www.smartsantander.eu/.
[24] http://www.smartproducts-project.eu/.

2

Resource Management Support of Big Data Procurement

Albena Mihovska and Sofoklis Kyriazakos
Center for TeleInfrastruktur (CTIF), Aalborg University, Aalborg, Denmark

John Soldatos
Athens Information Technology, Athens, Greece

2.1 Introduction

The Future Internet sets out a new vision for using cloud computing capacities for provision and support of ubiquitous connectivity and real-time applications and services for smart cities' needs. Data procured from highly distributed, heterogeneous, real and virtual devices (sensors, actuators, smart devices) will be automatically managed, analyzed and controlled by distributed cloud-based services.

Cloud computing and Internet of Things (IoT) are nowadays two of the most prominent and popular ICT paradigms that are expected to shape the next era of computing. The cloud computing paradigm (McFedries 2008) realizes and promotes the delivery of hardware and software resources over the Internet and according to an on-demand utility based model. Depending on the type of computing resources delivered via the cloud, cloud services take different forms such as Network as a Service (NaaS), Infrastructure as a service (IaaS), Platform as a service (PaaS), Software as a service (SaaS), Storage as a service (STaaS) and more. These services hold to promise to deliver increased reliability, security, high availability and improved QoS at an overall lower total cost of ownership (TCO).

At the same time, the IoT paradigm relies on the identification and use of a large number of heterogeneous physical and virtual objects (i.e. both physical and virtual representations), which are connected to the Internet (Vermesan 2011). IoT enables the communication between different objects,

Vladimir Poulkov and Ramjee Prasad (Eds.), Resource Management in Future Internet, 21–38.

as well as the in-context invocation of their capabilities (services) towards added-value applications. Early IoT applications are based on RFID (Radio Frequency Identification) and Wireless Sensor Network (WSN) technologies and deliver tangible benefits in several areas including manufacturing, logistics, trade, retail, green/sustainable applications, as well as other sectors.

This Chapter focuses on the challenges in current cloud resource management to support big data and the need for the decentralization of the related mechanisms. The Chapter is further organized as follows. Section 2.2 explains the importance of converging the IoT and cloud computing areas and gives a state of the art of the efforts made until today to enable this convergence. The challenges are explained and possible solutions for overcoming those are outlined and motivated. The remaining open issues are also rationalized with recommendations for the possible directions to be undertaken. Section 2.3 concludes the chapter.

2.2 Rationale for IoT - Cloud Convergence – Early Convergence Efforts

Since the early instantiations/implementations of both technologies, it has become apparent that their convergence could lead to a range of multiplicative benefits. Most IoT applications entail a large number of heterogeneous geographically distributed sensors. As a result, they need to handle many hundreds (sometimes thousands) of sensor streams, and could directly benefit from the immense distributed storage capacities of cloud computing infrastructures. Furthermore, cloud infrastructures could boost the computational capacities of IoT applications, given that several multi-sensor applications need to perform complex processing that is subject to timing (and other QoS constraints). Also, several IoT services (e.g., large scale sensing experiments, smart city applications) could benefit from a utility-based delivery paradigm, which emphasizes the on-demand establishment and delivery of IoT applications over a cloud-based infrastructure.

The proclaimed benefits of the IoT/cloud convergence have (early on) given rise to research efforts that attempted to integrate multi-sensory services into cloud computing infrastructures (Lim 2005). The most prominent of these efforts are the so-called sensor-clouds, which blend sensors into the data center of the cloud and accordingly provide service oriented access to sensor data and resources (Hassan 2009). Several recent research initiatives are focusing on real-life implementation of sensor clouds, including open source implementations. Note that such initiatives are in progress both in the US (Open Source IoT Cloud 2012) and in the EU (Open IoT Project 2012), aiming at developing middleware infrastructures for sensor-clouds, which will enable the on-demand delivery of IoT services.

In addition to research efforts towards open source sensor-clouds, there are also a large number of commercial on-line cloud-like infrastructures, which enable end-users to attach their devices on the cloud, while also enabling the development of applications that use those devices and the relevant sensor streams. Such commercial systems include COSM (cosm.com), Things Speak (thingspeak.com), and Sensor-Cloud (www.sensor-cloud.com). These systems provide tools for application development, but they offer very poor semantics and no readily available capabilities for utility based delivery. There is also a number of other projects which have been using cloud infrastructures as a medium for Machine-to-Machine (M2M) interactions (e.g., (Kranz 2010)), without however adapting the cloud infrastructure to the needs of the IoT delivery.

In the area of IoT applications (e.g., for smart cities), we are also witnessing cases of IoT/cloud convergence. For example, in the scope of the EU-funded ICT-PSP project RADICAL (RADICAL PSP], the partners will be deploying sensor streams over the BONFIRE cloud infrastructure proposed by the EU-funded ICT project BONFIRE (BONFIRE ICT), as means to benefitting from the cloud's storage capacity and applications hosting capabilities. A similar approach is followed in the scope of the EU-funded under FP7 Smart Santander (Smart Santander). Note, that smart cities is a privileged domain for exploring and realizing the IoT/cloud convergence, given that such applications need to manage and exploit a large number of distributed heterogeneous sensor streams and actuating devices.

2.2.1 Challenges Towards "True" Convergence

2.2.1.1 Requirements for the IoT and the cloud

IoT enables the collection, enrichment and distribution of a wide variety of data. A common aspect of IoT applications is that many of them of practical interest involve control and monitoring functions, where human-in-the-loop actions are not required. More and more Internet-enabled devices are at work to improve the productivity and quality of life of human beings, in all technological domains and their penetration is expected to grow exponentially in the next years [Santucci09].

The use of the Internet Protocol (IP), together with ubiquitous networks and cloud computing now allows any device to be equipped with a communications module. This enables devices to communicate status and information, which in turn can be aggregated, enriched and communicated internally or onwards to other units, thus allowing the use of the available data in new and useful ways. An example is to use the data gathered by the on-board computer of an automobile as part of the traction control system to tell cities where the roads are slippery [OECD12]. In essence, this concept is not new, but with the advent of IoT communication it can now allow for the data to be communicated to others, as well as further combined and enriched. With IoT and its subset machine-to-machine (M2M) communications [ITUM2M], an unprecedented opportunity is present to create countless applications and services that go far beyond the mere purpose of each participant. IoT together with its subset M2M, create a bridge between the real world (made of sensors, actuators, tags that are pervasive in our lives) and the virtual world (the Internet and its associated services).

Some of the data related to the IoT applications and services, will be generated by the public sector and will be of use to the general public. Governments are strong initiator of IoT use and large-scale IoT users themselves [OECD12]. Other data will be generated by private IoT-users and will be of use to public organisations. Adequate arrangements to give widespread use of data are constantly encouraged. Although, the cloud can offer Network, Infrastructures, Platforms or Software as a Service to a variety of end user (customer) requests, namely, storage of sensing data, computing of analytical data, data mining and processing, visualization of data, and so forth, the main challenge for its successful adoption for real-time IoT data delivery remains in the fact, that IoT services and data can be widely dispersed and fine-grained. IoT services might provide streaming data, and yet in some situations data communications could be of low quality and reliability, due to inherent resource and performance constraints. IoT services are very much inferred by the occurrence of a particular situation, and a lot of personalized significance can be associated to the obtained data. Interpretation and aggregation of raw sensor data is usually done based on a learning process. A full overview of IoT services and related requirements are mapped in Table 2.1.

The cloud infrastructure includes servers, storage, networks, and other hardware. It can deliver infrastructure resources as a service. Virtualization allows the splitting of a single physical piece of hardware into independent, self-governed environments, which can be extended in terms of CPU, RAM, Disk, I/O and other elements.

Table 2.1 IoT Service Characteristics Mapped to IoT Requirements

IoT Service Characteristics	IoT Requirements
• Data quality of various sources - Accuracy of each data point - Sensor reliability and availability - Time of measurement • Granularity of devices and data (e.g., real sensors, virtual sensors) • Geo-location • Streaming data • Resource constraints & performance requirements • Reliability, Quality of Information, Quality of Service	**Generic**: • Identification-based connectivity: based on a unique identifier Interoperability • Scalability: handling of large amounts of things and data • Autonomous services for data transferring and information exchange b/n things • Autonomous services for querying and automatically informing status of things • Autonomous services for collaboration among things • Automatic representation and processing of semantics of things • Control and management of things • Open APIs for thing related services • Self-Management (e.g., access, fault, configuration) **Service and Application related:** • Autonomic services provisioning • Location-based requirements: sense and track the location information automatically • High quality and highly secure human body related services • Customization of services based on business needs

The common characteristics of cloud infrastructure include the following specifics:

- Network centric: The framework of cloud infrastructure consists of plenty of computing resource, storage resource, and other hardware devices that connect with each other through a network.
- Service provisioning: Cloud infrastructure provides a multi-level on-demand service mode according to the individualized demand of different customers.
- High scalability/reliability: The provisioned resources are usually redundant, with backups stored for access in case of failure of the current working resources.
- Resource pooling/transparency: The underlying resources (computing, storage, network, etc.) of the cloud infrastructure are transparent to the

customer; the customer does not need to know how and where resources are deployed.

- In order for a cloud system to support the flow of real-time data for IoT applications, the following general cloud requirements can be identified.
- Related to data rates and latency: The data should be pushed out to the cloud, and then pushed to the user because polling requires too much time and uses too much bandwidth. Further, the data needs to flow quickly and effortlessly through the system, through a real-time database, which means that it should be able to stay in its simplest format.
- Related to security: the IoT application should be able to stream its data into the cloud without exposing its data to the dangers of communicating it over the Internet. Further, support of different user types accessing a single service should also be enabled. Resource configurations should also be performed in a secure way for each service across multiple cloud infrastructures.
- Related to semantics: the information provided by different sensors or other IoT devices is usually in a specific format allowing it only to be used by a proprietary application or to be controlled only by specific systems. But the provided information could be used by many other applications. It is therefore necessary for open interfaces and data formats to allow for using the different sensor outputs in applications, programming etc.
- Related to mobility: Specific and smart mechanisms should exist for controlling and exploiting the mobility of real-world entities and attached IoT devices.

The key requirements for the cloud that must be met to handle IoT data and services in real-time are summarized in Table 2.2.

To overcome the above limitations and challenges, it is proposed to realize the interoperable decentralized service-oriented cloud platform based on an open and context-centric approach.

2.2.1.2 Realization of a converged IoT/Cloud framework

To realize a "true" IoT/cloud convergence, a holistic approach aiming to optimize the cloud for IoT applications and vice versa, is necessary. In the current state of the art, the common approach is to interconnect sensor streams and IoT services with existing cloud middleware, rather than changing the cloud middleware itself towards optimizing the cloud for IoT services delivery. As a result, no true blending of the IoT and cloud computing paradigms is accomplished, which calls for considerable changes in the core

Table 2.2 Summary of IoT and real-time cloud requirements

IoT Requirements	Real-Time Cloud Requirements
Highly dynamic resource demand	Application elasticity
Real-time needs	Quality of service assurance
Expected exponential growth on demand	Cloud infrastructure scalability
Availability of applications	Cloud reliability
Data protection and user privacy	Cloud privacy and security
Efficient power consumption of applications	Efficient energy resource management
Execution of the applications near to end users	Inter-cloud operation; interoperability and portability

of cloud infrastructures. Such radical changes are required given the radical differences and conflicting properties of the two technologies, in particular (Lee 2010):

- IoT elements and infrastructures (e.g., sensors, WSN, RFID) are location specific, resource constrained and usually expensive to be developed and deployed. Consequently, IoT infrastructures are in general inflexible in terms of resource access and availability.
- On the contrary, cloud-based infrastructures are location independent and provide a wealth of easily accessible and usually inexpensive resources. Furthermore, cloud computing infrastructures are characterized by rapid elasticity, which is the foundation of their immense capacity and cost effectiveness.

These differences have a direct impact on how other aspects of Cloud/IoT convergence may be implemented, including virtualization, resource management, visualization, security and privacy. In particular:

- **Virtualization:** Virtualization in cloud computing is straightforward as a result of the location-independence of the resources, while it is much more challenging when it comes to dealing with location-dependence sensors/devices.
- **Resource Management**: The blending of IoT resources into the cloud introduces new resource management requirements, which are associated with the need to optimize not only processing, storage and I/O resources, but also sensor reading cycles, multi-sensor queries and more general shared access to expensive location-dependent IoT resources.
- **Privacy Management**: Current cloud computing infrastructures do not take any provisions for privacy protection associated with the collection

and tracking of personal data. There should be reconsidered, given that sensors/IoT devices make it possible to aggregate such data. Hence, cloud middleware needs to be revised in order to provide inherent privacy protection capabilities.

- **Visualization**: The visualization of IoT services is associated with challenges that do not exist in conventional transaction cloud computing services. The challenges relate for example to the need for real-time interactions with both physical and virtual internet-connected objects, which give rise to virtual reality (VR) interactions between end-users, the physical and the virtual world.
- **Real-Time Capabilities**: The involvement of multiple physical sensors in the scope of service delivery creates additional challenges associated with real-time interactions, which imposes a need for studying extensions over real-time operating systems (such as Free RTOS) for embedded devices, as well as how they could be supported in the scope of a cloud environment.

One of the biggest technological challenges at the moment is how cloud computing is applicable to real-time data for large-scale industrial and embedded applications (i.e., big data). This challenge opens new horizons in the highly demanding IoT world enriched and completed by the cloud paradigm and approach. The authors of this chapter claim that this challenge can be overcome by decentralizing the concept of the current cloud operation for answering the demands of the distributed in nature IoT applications. The main technological challenge is to integrate successfully the currently centralized concept of the cloud and its utilities to the highly distributed type of platforms on which IoT applications and services are based. An integrated combined framework utilizing the computing and storage capacity of the cloud to all ends of IoT communications and services can become a powerful tool to build new businesses.

A conceptual framework of integrating the cloud for the delivery of real-time data coming from ubiquitous (i.e. distributed) sensing and other similar devices constituting the IoT is shown in Figure 2.1.

In Figure 2.1, an open sensor platform with interfaces for the things and API for sensing and actuation is integrated with a distributed cloud-service platform for the automatic management, analysis and control of procured big data from large-scale real and virtual devices (e.g., sensor, actuators, etc). Both parts are closely coupled to result in a scalable platform enabling to share real-world data among heterogeneous devices and successfully integrating

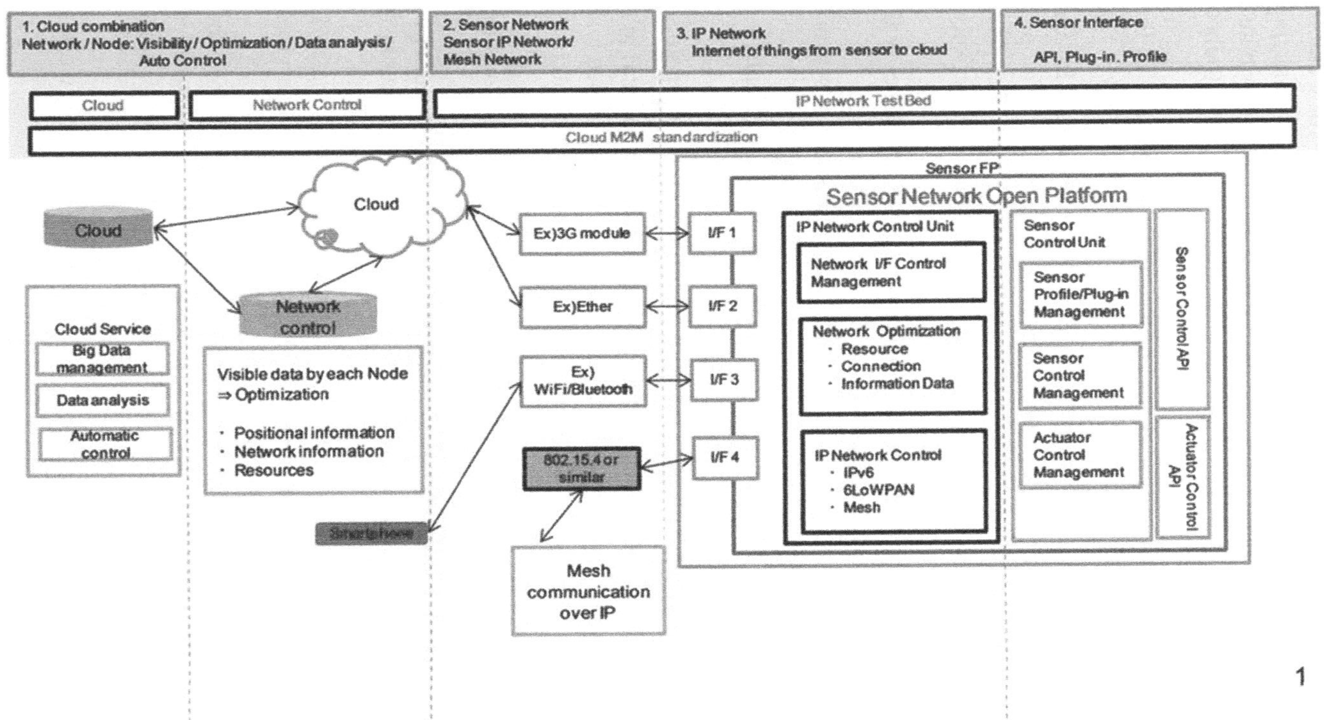

Figure 2.1 Conceptual representation of a converged IoT/cloud framework.

functionalities for cloud service and network control that provide for scalability, flexibility and on-demand resource provisioning in support of managing big size real-time data of various type and granularity; in highly distributed locations; and originating from resource- and computationally constrained devices.

The following five research issues for the successful convergence of the IoT and the cloud, arise from the concept shown in Figure 2.1, namely:

1. Cloud combination of an integrated network and service control for the automatic management, analysis and control of procured from an open sensor network big data.
2. Network management for the optimization of traffic; visibility between cloud datacenters and interoperability.
3. Sensor IP network where mesh communications may be realized over IP and may use a short-range air interface.
4. IP network construction for the seamless transport of data from sensors to the cloud and the reliable operation of the sensor network .
5. Sensor control for the easy, reliable and flexible operation and management of the sensor network data.

The converged framework should implement the following characteristics:

- Ability to virtualize location dependent entities, such as sensors and IoT services. To this end, the cloud computing platform will introduce and implement the concept of equivalent/similar physical and virtual internet-connected objects.
- Privacy friendliness on the basis of inherence capabilities for handling and transferring private data.
- Support for real-time interactions. This support will take into account not only the capabilities/limitations of real-time operating systems, but also the capabilities of embedded devices involved in the IoT paradigm.
- Real-time visualization of IoT services, on the basis of models that will enable ambient multimodal interactivity between end-users, IoT services the devices that the latter comprise.
- Self-optimization of the cloud infrastructure not only in terms of the cloud computing resources, but also in terms of the IoT resources (sensor streams, sensor I/O cycles, distributed multi-sensor queries) that it comprises.

In pursuit of these capabilities, concepts and techniques from both worlds (IoT, cloud) must be implemented:

- Service descriptions and utility measurement specifications for IoT services, which will be executed over cloud infrastructures.
- Directory services that consider both conventional computational resources and IoT resources on the basis of appropriate semantics (ontologies).
- Scheduling and resource management techniques that take into account shared (multi-tenant) access to (heterogeneous location-dependent) IoT resources, in addition to conventional (location-independent) computing resources.
- Security and access controls mechanisms that take into account the sharing of both computing resources and IoT resources (e.g., physical and virtual sensors).

By realizing a converged platform with the above characteristics, one would be able to deliver a number of smart applications; for example, exploiting large numbers of devices and social networks with a view to enhancing energy efficiency and environmental effectiveness within a city.

2.2.1.3 Open research challenges

The following open research issues can be summarized as the ones critical for a successful IoT/cloud integration

- Need to introduce plug and play sensor capabilities, and control of accuracy of data measurement, maintenance and calibration status for the reliable and easy sensor data collection
- Need for a flexible, agile, and energy-efficient supporting sensor architecture, where nodes can be added and removed without disturbing the data communication process, and where power operating nodes use renewable energy sources (e.g., solar, wind, etc).
- Guaranteed stable wireless and sensor communication for power saving equipment under the available installation environment by building an IP mesh communication network based on an appropriate (e.g., 802.14.5 g/e) air interface.
- Novel protocols and methods able to assign any IP address to any virtual machine with support of IPv6 address allocation, configuration, addressing, data packet parity, and other IPv6 protocol stack functions.
- Novel solutions for a local area distributed system to provide reliable high-speed direct (point-to-point) communications, congestion-free links and uniform bandwidth between servers.

- Novel traffic randomization algorithms for ensuring maximized use of the available links.
- Novel service optimization techniques for an improved application performance.
- Cost-effective policy-based routing schemes to ensure quick and effortless flow of data through the system.
- Techniques for the interworking between the networking and application layers of the cloud system.
- Novel traffic models to guarantee the various requirements of latency, packet loss sensitivity and bandwidth of the big heterogeneous data.
- Service-aware topology policies to distinguish between traffic and how it should be flown through the network.
- Solutions for a unified protocol framework for communications within datacenters to enable interworking and migration in the cloud.
- Techniques for on-demand resource provisioning.
- Security measures to ensure the protection and confidentiality of the procured data.

In order to realize real-time cloud services, the functional components of the converged IoT/cloud framework must answer the requirements of flexibility, scalability, on-demand resource provisioning and offer the necessary advanced network functions to ensure performance, security and availability of the cloud services. Further, support of decentralized operation must also be ensured.

Pooling and virtualization of physical resources are essential means to achieve the on-demand and elastic characteristics of the cloud. Through these processes, physical resources are turned into virtual machines, virtual storages, and virtual networks. These virtual resources are in turn managed and controlled by the Network/Resource Orchestration based on user demand.

The demand for real-time management of big data in support of IoT applications invokes the following minimum requirement characteristics of a converged IoT/cloud framework:

- **Scalability**: There are a lot of limitations of current technologies to provide large scale Ethernet domains. Some of these limitations are Address Resolution Protocol (ARP) broadcast limitations, MAC size tables constraints and Spanning Tree Protocol limitations. A lot of initiatives within the industry and the academia are proposed to solve these issues. An example is the Transport Interconnection of a Lot of Links (TRILL) protocol under standardization at the IETF [RFC6325, RFC6361]. Other possible solutions for separating names from locations are similar to the Locator/ID Separation Protocol (LISP) under standardization at the IETF. Scalability can be guaranteed by novel protocols and methods able to assign any IP address to any virtual machine and thus breaking all the scalability constraints of current IP sub-networking. Here, seamless migration of virtual resources without the need to change the IP address or any other open connection should be considered. The virtualization layer of the cloud platform should support IPv6 address allocation, configuration, addressing, data packet parity, and other IPv6 protocol stack functions.

- **Performance:** The nature of the cloud generated traffic is very random [Kandula09]. From an architecture perspective, the current 3-tier topology (access, aggregation and core) used in datacenters is not well adapted to provide reliable high-speed direct (point-to-point) communications between servers and congestion-free links and uniform bandwidth between any two arbitrary servers within a datacenter. To support these requirements, novel solutions for a local area distributed system as well as novel algorithms on how to maximize the use of the available links is needed. Existing research initiatives on traffic randomization were reported in [Niranjan09] to maximize the utilization of available links with 802.1ah or IP in IP encapsulation. Novel service optimization techniques should ensure the required application performance.
- **Agility and flexibility**: The cloud platform should be able to follow the high dynamics of cloud resources and to deliver new cloud services to ensure quick and effortless flow of data through the system. For example, the system should adapt quickly to a changing topology due to the mobility of virtual machines. Several propositions are under standardization to enable this virtual-machine aware networking (e.g., VEPA and VN-tag standard initiatives of IEEE). Solutions to have a fine-grained control of flows routing within the datacenter together with policy-based routing schemes can support the real-time data flows. Current solutions to do this are very costly (based on dedicated VLANs and Access Lists ACLs).
- **Network intelligence and proximity services:** To answer the demands of multidirectional traffic, the network layers need to provide enhanced visibility into the cloud datacenter services. There are several existing techniques which determine and implement the proximity services in the network, such as Domain Name System (DNS), Round-Trip Time (RTT) measurements, and manual configuration. However, these approaches do not provide interworking between the network layer and

application layers, which brings inflexibility and unreliability. For this, capabilities/interfaces to optimize application traffic, which are based on accurate information like management statistics, and policy database are needed.

- **QoS**: With big data traffic coming from trillions of sources, there will be much more traffic between datacenters. In a traditional implementation, all the traffic between data centers must run through the backbone, which puts a great burden on it. Further, the traffic models should be varied with applications, such as data synchronous, huge data replication, transmission relay of high-resolution video and data floating of frequently used services, etc. In addition, different datacenters may have respective workloads. For instance, there could be transaction-related, data processing related or storage-related workload. They require different traffic models to guarantee various requirements of latency, packet loss sensitivity and bandwidth. A requirement for the inter-cloud model is to provide for high degrees of agility of the architecture. To ensure that the backbone burden is minimized, service-aware topology policies should be implemented to distinguish between traffic and how it should be flown through the network. In addition to being able to route packets effectively and efficiently, the topology policies should allow for adjustments to the specific requirements. This ability to change topologies can be driven through an API so that agents can deal with the required changes without human intervention.
- **Reducing the Total Cost of Ownership (TCO):** There are many use cases where interconnecting cloud datacenters is required (recovery, maintenance, resources optimization, hybrid clouds). Layer 3 datacenters interconnection does not raise important challenges, there are well proven technologies that have been widely used and are in continual enhancements to answer any new requirements. On the other hand, there is an emerging technical challenge when choosing the technology to use for providing layer 2 extensions across geographically distributed datacenters to interconnect cloud infrastructures at layer 2. This is for example required to perform live Virtual Machines migration between data centers. Existing layer 2 solutions have several limitations that make them insufficient to fulfill advanced requirements in term of flexibility, scalability and availability of cloud computing services. Such challenges may be solved by unified protocol framework for communications within datacenters to satisfy the above requirements and at the same time offer a reduction in the TCO.

- **On demand resource provisioning**: The capability to describe all kinds of resources, including computing, storage and network resources and to support both virtual and physical resources is essential for the cloud. For example, resource packaging management is required to provide a unified interface of heterogeneous resources, whether they are virtual or physical. Resources can be managed hierarchically to satisfy traditional applications/services or enterprise environment hosting. Cloud resource management is required to be able to evaluate the performance of each resource to fulfill the QoS of each user request. A unified resource (including virtualized and physical computing resource, storage resource, and network resource) management function for the upper-layers. The resource management function should provide resource packaging, resource deployment, resource scheduling, whilst managing templates and assets. All resources should be flexible, on-demand and automation orchestrated, deployed and provisioned based on the pre-defined policies, which include high available, load balance, resources migration, energy efficient and storage deployment. It should be possible for the services to be analyzed and to be translated into resource requirements and to trigger appropriate actions. Resource scheduling algorithms that dynamically allocate resources by the real-time monitoring of applications and Service Level Agreements (SLAs) are a must here. Monitoring is essential to ensure the availability, security and usability of cloud services. Cloud infrastructure environments consist of a number of physical and virtual entities such as physical servers, virtual machine monitors/hypervisors, VMs, physical and virtual disks, physical and virtual network, and applications. All of these elements are associated through complex relationships. Data is captured for analysis and measurement by monitoring various attributes of these elements and their relationships. Real-time monitoring to be executed for all physical and virtual resources is a requirement for support of data management. The architecture of the resource monitor should be multi-layered, which includes service instance monitoring, physical resources monitoring, resource pool monitoring, user connection monitoring, software monitoring etc. It would be possible to detect the exception or error of computing, storage and network equipment and the resources pool while the currently connected number and the users' IP address, login time, idle time etc are monitored.

- **Security**: The cloud resource management is responsible for the entire process of resource allocation, authentication, accounting and related security in the cloud.

2.3 Conclusions

For a systematic solution to support service engineering for a wide range of scenarios that blend cloud computing with the internet-of-things (i.e. resulting in the so-called cloud of things) a more holistic and integrated approach for building cloud applications that involve sensors and other devices is needed.

The current service engineering paradigm of IoT can be advanced on the basis of tools and techniques for creating, configuring and orchestrating device-associated services that operate through the cloud and/or within the cloud. Enhancing hybrid approaches for service composition so that they can accommodate large volumes of heterogeneous sensor and device networks is one direction to undertake. The main idea is actually to develop new decomposition techniques that partition a set of end-to-end constraints into a collection of local ones. The latter will then be satisfied by independent, self-organized groups of services in a decentralized way. To support the seamless and dynamic integration of heterogeneous devices in these groups, each component will be modelled as an autonomous, intelligent agent that interacts locally with its peers. Following the principles of Swarm Intelligence, the aggregate behavior of local groups of agents is expected to exhibit higher intelligence than the individual components, ensuring sufficiently intelligent behavior for solving the local optimization problem assigned to them. Special care will also be taken to ensure high levels of fault-tolerance inside and across agent groups, while achieving high responsiveness (i.e., real-time behaviour). The latter can be actually ensured by the cloud infrastructure, which allows for parallelizing the activity (i.e., communication) inside each agent group.

References

[1] [McFedries2008]: P. McFedries, "The Cloud is the Computer," in IEEE Spectrum Magazine, August 2008.

[2] [Vermesan 211]: Ovidiu Vermesan and Peter Friess , "Internet of Things - Global Technological and Societal Trends," River Publishers, 2011, ISBN 9788792329677.

[3] [Lim05] H.B. Lim, et al. Sensor Grid: Integration of Wireless Sensor Networks and the Grid, In Proc. of the IEEE Conference on Local Computer Networks 30th Anniversary (LCN'05), October 2005.

[4] Hassan09] Mohammad Mehedi Hassan, Biao Song, Eui-nam Huh: A framework of sensor-cloud integration opportunities and challenges. ICUIMC 2009: 618–626.

[5] [Open Source IoT Cloud 2012]: https://github.com/iotcloud/IoTCloud.

[6] [OPEN IoT Project 2012] EU-funded FP7 project Open Source Solution for the Internet of Things into the Cloud, http://openiot.eu/?q=node/1.

[7] [Kranz 2010]: M.s Kranz, L. Roalter, and Fl. Michahelles, "Things That Twitter: Social Networks and the Internet of Things", What can the Internet of Things do for the Citizen (CIoT) Workshop at The Eighth International Conference on Pervasive Computing (Pervasive 2010), Helsinki, Finland, May 2010.

[8] [RADICAL PSP] EU-funded ICT-PSP project RADICAL: http://cordis.europa.eu/fp7/ict/fire/connected-smart-cities/presentations/17-radical.pdf.

[9] [BONFIRE ICT] EU-funded ICT Project BONFIRE: www.bonfire-project.eu/infrastructure.

[10] [SmartSantander ICT]: EU-funded FP7 project SmartSantander, www.smartsantander.eu/.

[11] [Santucci09] G. Santucci, "Smart networks, objects, buildings and people: Empowering the Internet for Smarter Cities", European Commission, November 2009.

[12] [OECD12] OECD (2012), "Machine-to-Machine Communications: Connecting Billions of Devices", OECD Digital Economy Papers, No. 192, OECD Publishing. http://dx.doi.org/10.1787/5k9gsh2gp043-en.

[13] [ITUM2M] ITU-T Standardization, Focus Group on M2M, http://www.itu.int/en/ITU-T/focusgroups/m2m/Pages/default.aspx.

[14] [Lee 2010] K. Lee and D. Hughes, "System architecture directions for tangible cloud computing," In International Workshop on Information Security and Applications (IWISA 2010), in Qinhuangdao, China, October 22–25, 2010.

[15] [RFC6325] Internet Engineering Task Force (IETF); RFC 6325, "Routing Bridges (RBridges): Base Protocol Specification," July 2011, www.ietf.org.

[16] [RFC6361] Internet Engineering Task Force (IETF); RFC 6361, "PPP Transparent Interconnection of Lots of Links (TRILL) Protocol Control Protocol," August 2011, www.ietf.org.
[17] [Kandula09] Kandula, A., Sengupta, S., Patel, P. The nature of data center traffic: Measurements and analysis. In: ACM SIGCOMM IMC (November 2009).

3

Future Internet Technologies for Open Access to Resource Management in Multimedia Networks

Ivaylo Atanasov and Evelina Pencheva
Faculty of Telecommunications, Technical University of Sofia, Bulgaria
E-mail: Ivailo Atanasov< iia@tu-sofia.bg >;
Evelina Pencheva <enp@tu-sofia.bg>

3.1 Introduction

The convergence of Internet and telecommunications creates opportunities to deliver new attractive applications, using common technology, developing new business models, or through services being offered as packages. A lot of very different services can be offered on top of a network based on the Internet Protocol (IP). Different media such as voice, video, text and data may be encoded in digital form, and common approaches to storage, manipulation, co-ordination and transmission of all forms of information are possible using Future Internet technologies. A global, access independent service control architecture that is based on IP connectivity is IP Multimedia Subsystem (IMS). The IMS provides full integration of voice and data services increasing productivity and overall effectiveness. It combines mobility and internet which is the main prerequisite for service success in the future [10].

Convergence between information technology applications and public telecommunications services requires the network to be opened, allowing applications in an external domain to invoke functionality in a public network. There are several approaches to open the network and achieve resource programmability. One of them is the standardized Open Service Access (OSA) architecture.

In IMS, quality of service (QoS) is seen as one of the aspects that is used in order to differentiate multimedia offerings from those of common internet service providers. The IMS provides end-to-end QoS through efficient

Vladimir Poulkov and Ramjee Prasad (Eds.), Resource Management in Future Internet, 39–74.

mechanisms for resource management. The resource management includes authorization and monitoring of usage of network resources intended for multimedia traffic. The policy control enables the introduction of more intelligence in the resource management decisions for QoS. In IMS, the overall concept for policy based QoS control and flow based charging is called Policy and Charging Control (PCC) [4].

The chapter investigates capabilities for open access to resource management functions in all IP-based multimedia networks in the context of PCC. In such networks, session initiation, modification and termination rely on Session Initiation Protocol (SIP) signalling. Procedures related to authorization and usage monitoring of bearer resources are conveyed by the use of Diameter protocol. The open access to management of bearer resources intended for multimedia traffic allows third party applications to change dynamically the QoS parameters as defined in the user profile or available on established multimedia sessions. The access to resource management functions is through application programming interfaces (APIs) that provide high level abstraction of network functions and hide for application developers the complexity of control protocols. Deployment of third party resource management in IP-based multimedia networks requires mapping of APIs onto network control protocols such as SIP and Diameter.

The open access to telecommunication functions may be provided by OSA API [23]. The OSA Connectivity Manager API allows negotiation and management of QoS and service level agreements in IP networks. The usage of OSA APIs requires some expertise in telecommunications as the provided level of abstraction of network functions is low. Parlay X Web services are intended to provide open service access with a simpler interface and greater abstraction of the underlying network and data resources [8]. Parlay X Application-driven Quality of Service is a service which makes the available QoS on user connections in the network manageable.

A thorough analysis of the supported resource management functionality shows that currently the OSA APIs and Parlay X Web services do not support the functionality for open access to PCC. The chapter presents an approach to design of OSA compliant APIs for PCC.

The access of third party applications to specific network functions requires special type of application server to be deployed by the telecom operator. This application server needs to provide APIs toward applications and stacks of control protocols toward the network. It is responsible for translation of API methods invocations into control protocol messages and vice versa. In addition, it needs to maintain mutually synchronized state machines representing the application view on the state of QoS resources and the protocol view on the session state. The chapter investigates some implementation issues of such application server including interface to protocol mapping and functional verification.

When the network operator deploys an application server of any kind, one of the main issues is the server performance and utilization [14]. The open access is expected to increase the number of potential service developers that will access telecommunication services. The applications residing in internet can reach communication functions via application server. The chapter presents a model of application server for open service access, proposes a dynamic mechanism for admission control and evaluates application server utilization. The conclusion summarizes the contributions and highlights the benefits of the suggested approaches.

The chapter is structured as follows. The related works in the area are reviewed in Section 3.2 Section 3.3 presents the logical architecture for deployment of open access to resource management functions. PCC architecture with User Data Convergence is considered and the requirements to open access to PCC are identified. The standardized capabilities for open access to resource management are evaluated. Section 3.4 describes an approach to design of APIs for PCC. The designed interfaces are mapped onto Diameter protocol. Models of QoS resources intended for multimedia traffic are suggested. The models represent the application and the network views on QoS resources. An approach to functional verification of application server providing API for resource management and network protocol is suggested. Use cases that illustrate the capabilities for open access to resource management functions in all IP-based multimedia are described. Section 3.5 describes a model of special type of application server for open access to resource management – Parlay X gateway. The model considers traffic of different priorities generated by the service providers and by the network which has to be served. It takes into account the distributed architecture of the Parlay X gateway and applies mechanisms for adaptive admission control and load balancing to prevent overloading. The Parlay X gateway utilization is formalized mathematically. Simulations are used to evaluate the performance.

3.2 Related Works

PCC allows flexible quality of service (QoS) management of ongoing multimedia sessions in case of changing both the access networks and user devices with different capabilities. The PCC can also contribute to seamless service continuity in case of handover between two wireless networks without user intervention and with minimal service disruptions. Advanced solutions for policy-based QoS control are suggested in [17–18, 24].

Good & Ventura [11] propose a multilayered policy control architecture that extends the general resource management function being standardized. The extended architecture gives application developers greater control over the way the services are treated in the transport layer. Good, Gouveia, Ventura & Magedanz [13] suggest enhancements to the PCC framework that extend the end-to-end inter-domain mechanisms to discover the signalling routes at the service control layer, and use this to determine the paths traversed by the media at the resource control layer. Because the approach operates at these layers, it is compatible with existing transport networks and exploits already existing QoS control mechanisms. Policy-based service provisioning system is proposed [22] in order to provide different classes of services.

To stimulate service provisioning and to allow applications outside of network operator domain to invoke communication functions, an approach to opening the network interfaces is developed [12]. Stojanovic, Rakas & Acimovic-Raspopovic [21] address an open issue of end-to-end service specification and mapping in next generation networks. A centralized approach has been considered, via the third party agent that manages negotiation process in a group of domains. The authors suggest a general structure of the service specification form, which contains technical parameters related to a particular service request. Bormann, Braun, Flake & Tacken [4] extend the mediation layer between the operators' core network and the charging system by adding capabilities for online charging control. The authors present a prototype that implements and extends parts of the standardized PCC architecture. Akhatar [1] develops a system and method for providing QoS enablers for third party applications. In one embodiment, the method comprises user equipment establishing a session with a third party application server hosting a selected third party application and receiving from the third party application server QoS information comprising at least one of a plurality of QoS attributes and configuring a QoS of a radio access network in accordance with the obtained QoS information. The method further comprises activating the radio access network QoS for the selected application, and establishing an application session with the third party application server via the radio access network. In [7], the authors propose novel access reselection procedures which enable a network provider to optimize the allocation of the users on the different access networks available using as central concept that handovers can and should be triggered by the modifications on the resources required by the mobile devices in order to optimize the overall usage of the wireless environment.

OSA APIs, standardized by 3GPP, provide access to network functions upon which application developers can rely when designing new services or enhancements (versions) of already existing ones. OSA API for QoS management is called "Connectivity Manager" and it is defined in 3GPP TS 23.198–10. The API is used for negotiation and management of QoS and service level agreements in IP networks.

The Parlay X "Application-Driven Quality of Service" (ADQ), defined in 3GPP TS 29.199–17, allows applications to control the QoS available on user connection. It may be used for dynamic management of QoS parameters available on multimedia sessions. The Parlay X APIs are defined before the standardization of IMS PCC. The analysis on PCC functions shows that these interfaces do not cover all QoS management functions that network operator can expose.

Parlay X APIs enable creation of a SOA (Service Oriented Architecture) solution. A common problem in deployment of SOA applications is the server performance and utilization. SOA grids can be used to break the convention of stateless-only services for scalability and high availability by allowing stateful conversations to occur across multiple service requests [6]. Due to great advantages that SOA offers to its adopters in almost all fields, many studies tried to leverage it in cloud computing. These studies focused on enabling easy access and flexible management to underlying grid resources [20]. There are still challenges for traditional applications of message-oriented middleware for achieving high levels of quality of service (QoS) when sharing data between services over an enterprise service bus. In [9], the authors present an analytical framework to derive the response time and service availability of client/server based SOA and Pear-to-Pear (P2P) based SOA. The impact on the response time and service availability for varying load conditions and connectivity for both client/server and P2P SOA implementation is studied. The authors of [5] suggest a SOA server virtualization driven by the goal of reducing the total number of physical servers in an organization by consolidating multiple applications on shared servers. The expected benefits include more efficient server utilization. However, SOA combined with server virtualization may significantly increase risks such as saturation and service level agreement (SLA) violations.

In the chapter, a dynamic load control mechanism of SOA server (Parlay X gateway) is proposed which can improve the performance and increases its utilization.

3.3 Open Access to Policy and Charging Control

Before identification of functional requirements to open access to PCC, the functional architecture is addressed, including the functional entities and the relationships between them.

3.3.1 Architecture for Open Service Access to Resource Management

A possible deployment of open access to resource management functions in PCC architecture is shown in Figure 3.1.

Policy and Charging Control architecture is defined in 3GPP TS 23.203 specifications. The Policy and Charging Rule Function (PCRF) encompasses policy control decisions and flow based charging control functionalities. The Policy and Charging Enforcement Function (PCEF) includes service data flow detection, policy enforcement and flow based charging functions. It is located at the media gateway. The Online Charging System (OCS) performs online credit control functions. It is responsible to interact in real time with the user's account and to control or to monitor the charges related to service usage. Offline Charging System (OFCS) is responsible for charging process where

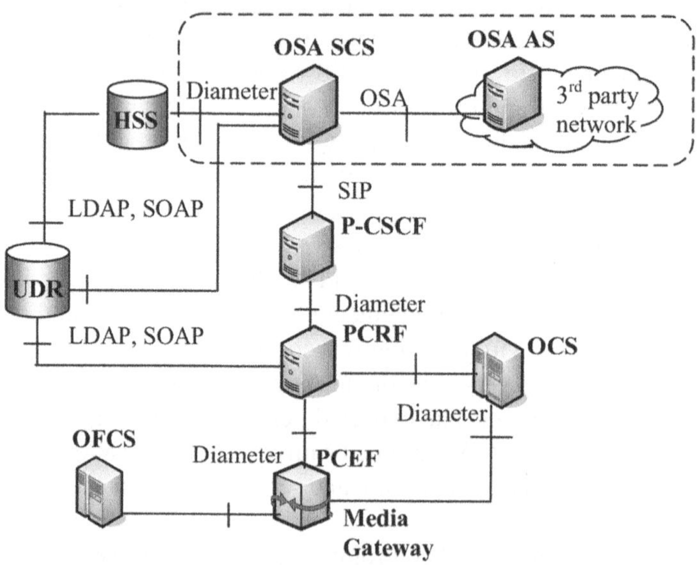

Figure 3.1 Logical architecture for deployment of open access to PCC.

charging information is mainly collected after the end of the session and it does not affect in real time the service being used.

The Home Subscriber Server (HSS) contains all subscription related information needed for PCC rules. If the PCC architecture supports User Data Convergence (UDC) defined in 3GPP TS 23.335 then the User Data Repository (UDR) acts as a single logical repository for user data. The user data may, for example, contain information about default QoS parameters which have to be applied each time the user creates a session. Functional entities such as HSS and Application Servers keep their application logic, but they do not locally store user data permanently.

Call Session Control Functions (CSCFs) include functions that are common for all services. The Proxy CSCF (P-CSCF) is the first point of contact for user equipment. It deals with SIP compression, secured routing of SIP messages and SIP sessions monitoring.

Application Servers (AS) run third party applications which are outside the network operator domain. OSA Service Capability Server (OSA SCS) is a special type of AS that provides APIs for third party applications and supports IMS protocols toward the network. The OSA SCS makes the translation between OSA APIs and control protocols.

The network operator may decide to provide the open access to resource management functions through Parlay X APIs by combining with existing OSA deployment configurations. In this scenario, the applications utilize Web Services to discover and interact with the network, without having visibility to the OSA implementation behind the Parlay X gateway. The Parlay X gateway interacts with the OSA SCS through OSA interface. The information published to the Web Services Registry provides the applications with the connection information required to connect with the Parlay X gateway. Alternative scenario for deployment of Parlay X Web Services is through Parlay X gateway which is connected to network node. The applications discover the Parlay X gateway and interact with the network without bothering of interface to protocol conversion which is responsibility of the Parlay X gateway. Both scenarios are presented in Figure 3.2.

Diameter is the control protocol in interfaces where authentication, authorization and accounting functions are required as to RFC 3588. The control protocol in interfaces where session management is performed is Session Initiation Protocol (SIP). Lightweight Data Access Protocol (LDAP) and Simple Object Access Protocol (SOAP) are the control protocols used to create, read, modify and delete user data in the UDR, and to subscribe to and receive notifications about user data changes.

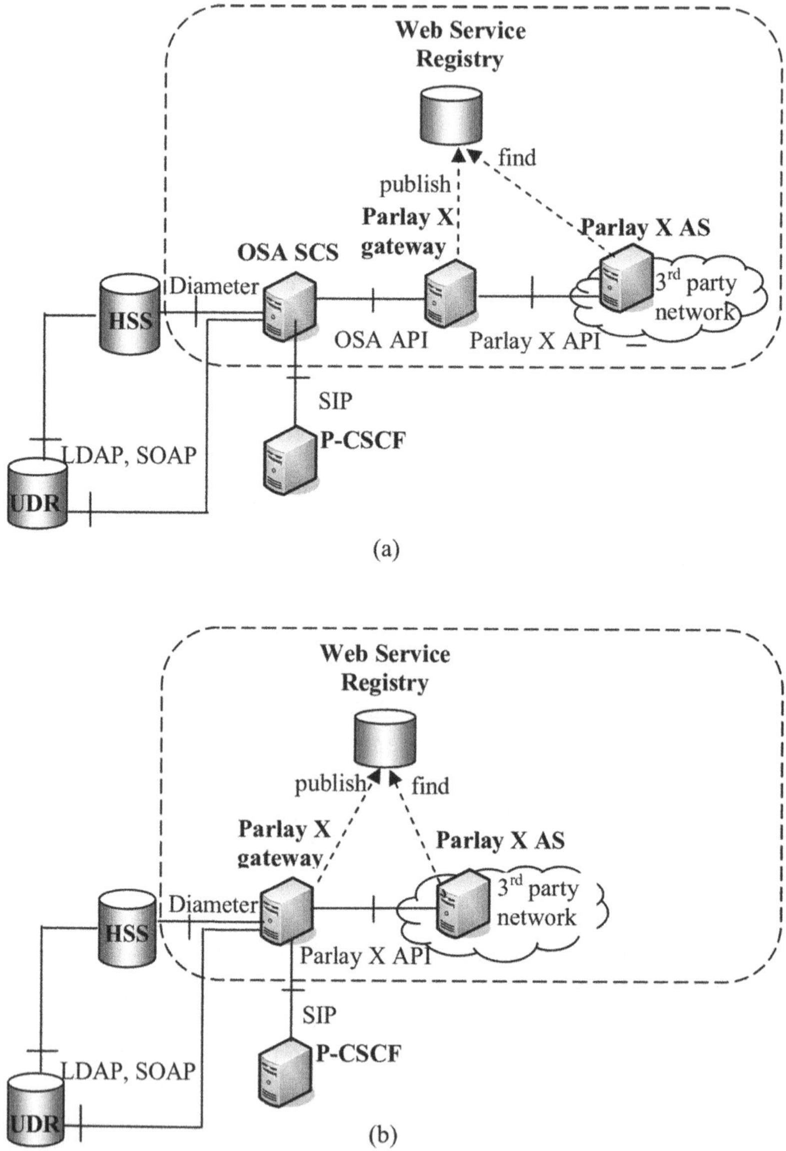

Figure 3.2 Deployment scenarios for Parlay X interfaces: (a) to OSA SCS; (b) to network elements directly.

Note that not all charging related interfaces and policy control functions are shown in Figure 3.1 for the sake of simplicity.

3.3.2 Identification of Requirements to Access to PCC

The PCC includes mechanisms for controlling the bearer traffic by using IP policies.

During the multimedia session establishment and modification, the user equipment negotiates a set of media characteristics. If the network operator applies policy control then the OSA SCS sends the relevant session description information to the PCRF in order to form IP QoS authorization data. The third party application can be involved in the process of QoS authorization by requesting specific QoS parameters to be applied, modified or removed.

Functional requirement 1: During the SIP session establishment, third party application may require to apply or to modify temporary specific QoS features on user session(s). The required functions include: to apply temporary QoS parameters, to modify temporary QoS parameters and/or to remove QoS parameters for a predefined duration (e.g. for session duration). The application logic is activated in case of session initiation, modification or termination.

In the Evolved Packet System (EPS), it is primary the network that decides what kind of bearer user equipment needs during communication. Having application/service information and based on subscription information and policies, PCRF provides its decision in a form of PCC rules which are used by the PCEF for gating control. Any QoS events, such as indication of bearer release or bearer loss/recovery, are reported by the PCEF to the PCRF and consequently to OSA SCS. Using the policy control capabilities, the OSA SCS is able to track the status of the IMS signalling and user plane bearers that the user equipment currently uses, and to receive notifications when some or all service data flows are deactivated.

To receive notifications about QoS events the third party application needs to manage its subscriptions for notifications. By using information about bearers and signalling path status the third party application can improve service execution. For example, the application can initiate session release on behalf of the user after indication that all service flows assigned to the ongoing session are released, but the P-CSCF has not received session termination request from the user equipment itself.

Functional requirement 2: The required functions for third party application in order to manage the QoS event subscription include the following: to create notifications and to set the criteria for QoS; to change notifications by modification of the QoS event criteria; to enable/disable notifications, and to query for the event criteria set; to report notifications upon QoS event occurrences.

Functional requirement 3: The third party application should be able to request QoS resource release. Using this function, the application can prevent unauthorized usage of bearer resources after SIP session termination.

The third party application may be interested in the accumulated usage of network resources on per session and user basis. This capability may be required for applying QoS control based on the total network usage in real-time. For example, the third party application may change the charging rate based on the resource usage (e.g. applying discounts after a specified volume have been reached). Another use is the assignment of a common quota for both fixed and mobile accesses for a limited time period for a defined set of subscriptions. During each session the network elements monitor the common quota which may be consumed by one or more devices over either the wireless or fixed networks. When a defined percentage of the common quota and/or all common quotas have been consumed, the third party application may be notified of the event. When the common quota has been consumed the third party application may prevent the access to the services.

Functional requirement 4: The third party application should be able to set the applicable thresholds for usage monitoring. Usage monitoring, if activated, shall be performed for a particular application, a group of applications or all detected traffic within a specific multimedia session. The third party application should be notified when the provided usage monitoring thresholds have been reached.

3.3.3 Evaluation of Standardized Capabilities for Open Access to PCC

The OSA "Connectivity Manager" API provides configuration of and control over the attributes of IP connectivity within and between IP domains. The APIs can be used to configure QoS parameters in a virtual private network (VPN) supported by IP networks. The VPN is provisioned using virtual leased line concept, termed virtual provisioned pipe (VPrP). Elements that can be specified for a VPrP include attributes such as packet delay and packet loss. Characteristics of traffic that enters the VPrP at its access point to the provider network can be also specified by attributes such as maximum rate and burst rate. The APIs allow QoS attributes to be set as default but do not provide means for assessment of the QoS provisioned. The APIs are more of operational control and cannot be used for dynamic QoS control on multimedia sessions.

Currently, there are no OSA APIs for dynamic QoS control and access to QoS related user data.

The "Application-Driven Quality of Service" (ADQ) is a Parlay X Web Service that allows applications to control the QoS available on user connection. Configurable service attributes are upstream rate, downstream rate and other QoS parameters specified by the service provider. Changes in the QoS may be applied either for defined time interval, or each time user connects to the network.

The ADQ ApplicationQoS interface defines operations for applying a new QoS feature to an end user connection. The ApplyQoSFeature operation is used by third party application to request a default QoS feature to be set up on the end user connection, which results in a permanent change in the class of service provided over the end user connection. A default QoS feature governs the traffic flow on the end user connection whenever there are no temporary QoS features active on the connection. The ApplyQoSFeature operation is used by third party application to request also a temporary QoS feature to be set up on the end user connection for a specified period of time. The ModifyQoS Feature operation is used by third party application to alter the configurable service attributes (e.g. duration) of an active temporary QoS feature instance. The RemoveQoSFeature operation is used by third party application to release a temporary QoS Feature, which is currently active on the end user connection. Therefore, these operations provide functions required to apply, modify and remove temporary QoS parameters (e.g. for session duration).

The ADQ Web Service enables applications to register with the service for notifications about network events that affect the QoS, temporary configured on the user's connection.

The ADQ Application QoS NotificationManager is used by third party application to manage their registration for notifications. The startQoSNotification operation is used by third party application to register their interest in receiving notifications of a specific event type(s) in context of specific end users. The stopQoSNotification operation is used by third party application to stop receiving notifications by canceling an existing registration. Therefore, these operations provide functions required to manage the QoS event subscription.

The ADQ ApplicationQoSNotification interface provides the operations for notifying the Application about the impact of certain events on QoS features that were active on the end user connection when these events occurred. The notifyQoS operation reports a network event that has occurred against end user(s) active QoS features. Therefore, this operation provides functions required to report notifications upon QoS event occurrence.

As to 3GPP TS 29.214 there are indications reported over the Rx reference point by the PCRF to the P-CSCF such as recovery of bearer, establishment of bearer, IP-CAN change, out of credit and usage report. These indications can not be forwarded to the third party application by the existing definition of the enumerated type QoSEvent.

Currently, not supported by ADQ Web Service functions required for policy control include usage monitoring and resources release.

The Parlay X "Application-Driven QoS" Web service defines operations which allow retrieval of the current status of user sessions, including history list of all QoS transactions previously requested against a user session. As far as the getQoSStatus operation of the ApplicationQoS interface is used by the third party application to access the currently available QoS features on a user session, it is impossible for third party application to retrieve the configured QoS features stored in the user profile. Further, if the QoS-related data in the user profile have been changed by administrative means, the third party application can not be notified.

The Parlay X "Payment" Web Service supports payment reservations, pre-paid payments, and post-paid payments. It supports payments for any content in an open, Web-like environment. When combined with ADQ Web Service, the "Payment" may be used for charging based on the negotiated QoS. The features for QoS based charging are restricted to temporary configured QoS parameters but can not reflect the dynamic QoS changes during the session. Flow-based charging is also impossible, as far as the Parlay X "Call notification" Web Service, defined in 3GPP TS 29.199–3, does not provide notifications about media addition or deletion for a particular session. Location based charging can be applied by combination of Parlay X "Terminal Location", defined in 3GPP TS 29.199–9, and "Payment" Web Services. Table 3.1 shows the Parlay X Web Services support for advanced charging.

Table 3.1 Advanced charging functions

Functions	Parlay X Web Services	Operations
QoS based charging	"Application-driven QoS" and "Payment"	notify QoS event and charge amount, refund amount
Time of day based charging	"Call notification" and "Payment"	notify called number and charge amount, refund amount
Location based charging	"Terminal location" and "Payment"	get location and charge amount, refund amount
Service flow based charging	"Audio call" and "Payment"	get media for participant and charge amount, refund amount

3.4 Design of Application Programing Interfaces to Policy and Charging Control

In this section an approach to definition of OSA compliant APIs for PCC based on identified requirements in Section 3.3.

3.4.1 Object-Oriented Interfaces for Quality of Service Control

The APIs provide a service named "Application-controlled Quality of Service". The network side interfaces with prefix Ip provide methods for PCC while the application side interfaces with prefix IpApp are used to receive results and notifications. The interfaces structure and dependences are shown in Figure 3.3. Below, the interfaces methods are described and mapped onto Diameter protocol considering their implementation in the network. Indexes are used to distinguish different content (Attribute Value Pairs) in the same Diameter command names.

The IpQoSManager interface inherits from OSA IpService and it is manager of the "Application-controlled Quality of Service". The interface provides functions for subscription management for QoS-related events. Using the methods of the IpQoSManager a third party application may define criteria for QoS related notifications, may enable or disable notifying and to query about or change the criteria for notifications. The createQoSResources() method may be used to create an object representing authorized QoS resources for given user session. A third party application may invoke the getQoSHistory Events() method to request tracing of dynamic changes related to QoS parameters.

The IpAppQoS Manager interface inherits from OSA IpInterface. It provides a method for QoS event notifications. It is used by a third party application to receive information about applied QoS parameters on user sessions and to receive a history list of QoS changes. The notifyQoSEvent() method reports events related to access network change, service data flow deactivation and signalling path status. It is mapped onto Diameter RAR command related to traffic plane events respectively (RAR_{IPCAN}, RAR_{SDFD}, RAR_{SPS}). The command is acknowledged by the respective Diameter RAA command (RAA_{IPCAN}, RAA_{SDFD}, RAA_{SPS}).

The IpQoSResources interface inherits from OSA IpService and provides methods for control of QoS resource authorization. Using the interface's methods a third party application may request specific QoS to be applied to user session, or modified and removed.

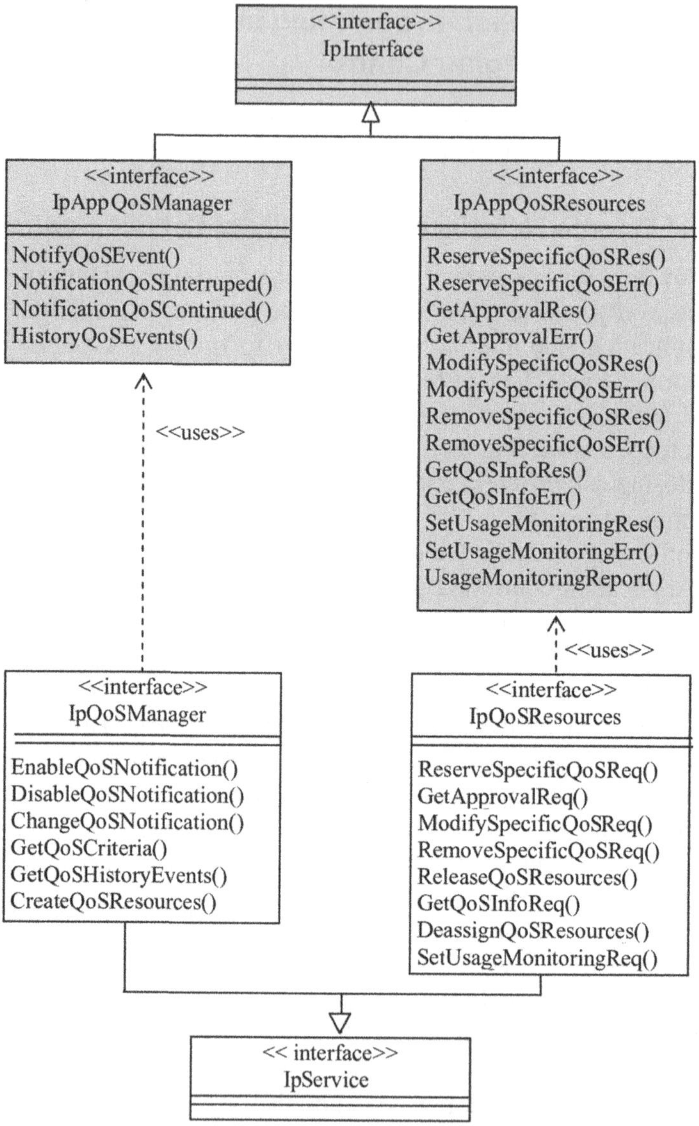

Figure 3.3 Application-controlled Quality of Service interfaces structure.

The Ip AppQoSResources interface is implemented by the application and inherits from OSA IpInterface. The interface methods are designed to receive results of the actions in the network required by a third party application using IpQoSResources.

The reserveSpecificQoSReq() method reserves specific QoS resources. It is mapped onto Diameter AAR (Authentication-Authorization Request) command related to initial provisioning of session information request (AAR_{IP}).

The Diameter command is acknowledged by AAA_{IP} (Authentication-Authorization Answer). The reserveSpecificQoSRes() method reports QoS resources reservation, and the reserveSpecificQoSErr() method reports QoS reservation errors.

The qosApprovalReq() method requests gate opening/closing. It is mapped onto Diameter AAR command related to gating control (AAR_{GC}) which is acknowledged by AAA_{GC}. The qosApprovalRes() method reports the result of gating control while the qosApprovalErr() method reports gating control errors.

The modifySpecificQoSReq() method modifies temporary QoS resources. It is mapped onto Diameter AAR command related to modification of session information request (AAR_{SM}) which is acknowledged by Diameter AAA_{SM} command. The modifySpecificQoSRes() method reports modification of QoS resources while the modifySpecificQoSErr() method reports QoS modification errors.

The removeSpecificQoSReq() method removes temporary set QoS resources. It is also mapped onto Diameter AAR command related to modification of session information request (AAR_{SM}). The command is acknowledged by AAA_{SM}. The removeSpecificQoSRes() method reports the result of QoS resources removal while the removeSpecificQoSErr() method reports QoS removal errors.

The releaseQoSResources() method requests release of QoS resources. It is mapped onto Diameter Session Termination Request (STR) command. The command is acknowledged by Session Termination Answer (STA) command.

The getQoSInfoReq() method requests QoS information. It is mapped onto Diameter AAR command related to initial provisioning of session information request (AAR_{IP}). The command is acknowledged by AAA_{IP}. The getQoSInfoRes() method provides the requested QoS information while the getQoSInfoErr() method reports errors.

The setUsageMonitoringReq() method is used to set usage monitoring thresholds. It is mapped onto Diameter AAR command related to usage monitoring (AAR_{UM}) which is acknowledged by AAA_{UM} command. The setUsageMonitoringRes() method reports results of settings monitoring thresholds while the setUsageMonitoringErr() method reports errors in setting monitoring thresholds.

The usageMonitoringReport() method reports events related to usage monitoring. It is mapped onto Diameter Re-Authorization Request (RAR) command related to usage monitoring (RAR_{UM}). The command is acknowledged by Re-Authorization Answer (RAA_{UM}) command.

The deassignQoSResources() method is used to release the relationship between the third party application and the QoS resources in the network.

3.4.2 Models of Quality of Service Resources

In addition to interface to protocol mapping, the OSA SCS has to maintain two state machines: one representing the application view on QoS resources and another one representing the Diameter session state. The state machines need to be synchronized i.e. to expose equivalent behaviour.

The QoSresources object reflects the application view on the status of QoS resources authorized for a user session. Figure 3.4 represents the states of the QoSresources object as seen by the application.

In order to simplify the description by PCCrequest it is denoted any of the methods modifySpecificQoSReq(), qosApprovalReq(), removeSpecific QoSReq() or set UsageMonitoringReq() and by PCCreport it is denoted any of

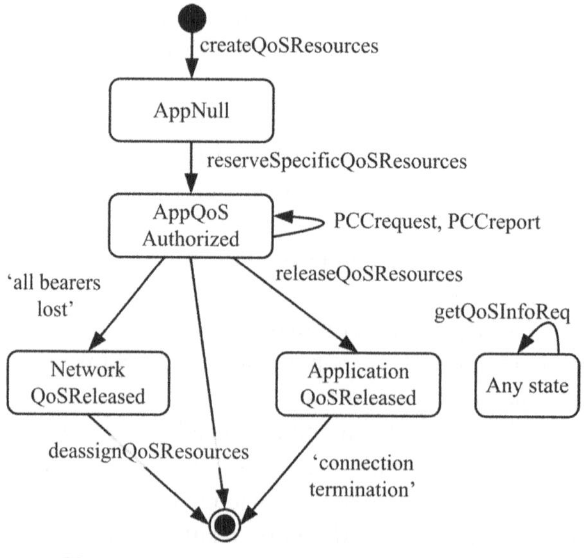

Figure 3.4 Application view on QoS resources.

notifyQoSEvent() method which provides QoS event notification excluding notifications about all bearers loss, and usageMonitoringReport() method.

The createQoSResources() method creates an instance of the QoS resources object. In AppNull state, there are no QoS resources authorized for the user session. In AppQoSAuthorized the QoS resources are authorized. In this state, the third party application may control the user traffic, may modify or remove temporary QoS features, and receive notifications about traffic plane events or reports on usage monitoring. The QoSresources object moves to ApplicationQoSReleased state when the third party application releases the QoSresources authorized for the user session. In this state, when the user session is terminated in the network, the QoSresources object moves to AppNull state. The QoSresources object moves to NetworkQoSReleased state when the third party application receives a notification that all bearers assigned to the user session are lost.

The mathematical notation of Labelled Transition Systems (LTS) is used to formalize the description of the application view on QoS resources. A *Labelled Transition System* (LTS) is a quadruple (S, Act, $\rightarrow$, s_0), where S is countable set of states, Act is a countable set of elementary actions, $\rightarrow \subseteq S \times Act \times S$ is a set of transitions, and $s_0 \in S$ is the set of initial states [19].

By $T_{ApPCC} = (S_{ApPCC}, Act_{ApPCC} \rightarrow_{ApPCC}, s'_0)$ it is denoted a LTS representing the application view on QoS resources where:

S_{ApPCC} = { AppNull, AppQoS Authorized, Application QoSRelease, NetworkQoSReleased};

Act_{ApPCC} = {reserveSpecificQoS, PCCrequest, PCCreports, allBearersLost, releaseQoSResources, connectionTermination, deassignQoSResources };

$\rightarrow_{ApPCC}$ = {AppNull reserveSpecificQoS AppQoSAuthorized, AppQoSAuthorized PCCrequest AppQoSAuthorized, AppQoSAuthorized PCCreportsAppQoSAuthorized, AppQoSAuthorized allBearersLost NetworkQoSReleased, AppQoSAuthorized releaseQoSResources ApplicationQoSReleased, NetworkQoSReleased deassign QoSResources AppNull, ApplicationQoSReleased connectionTermination AppNull};

s'_0 = {AppNull}.

A Diameter QoS resources session has to be described which represents the protocol view on QoS resources. The RFC 3588 defines Diameter peer state machines that describe the states of the Diameter peers in the dialogue. In an active Diameter dialogue, different Diameter sessions may run. The suggested Diameter QoS resources session is based on the Diameter application in IP Multimedia Subsystem for PCC procedures. The state machine of the Diameter QoS resources session is shown in Figure 3.5.

From PCC point of view, in Null state there are no QoS resources assigned to the user session. The initial provisioning of user session information moves the Diameter QoS resources session into QoSAuthorized state where QoS resources are assigned to the user session.

In the QoSAuthorized state, user session modification and gating control procedures may be executed. If the P-CSCF is subscribed to receive notifications, traffic plane events and usage monitoring events may be reported. The PCRF sends ASR command when the session in the access network providing IP connectivity is terminated. This triggers a transition to the Null state. When the third party application requests the user session termination, an STR command is sent to the PCRF.

By PCCprocedures it is denoted any of the Diameter commands AAR_{IP}, AAR_{SM}, AAR_{GC}, AAR_{SM}. By PCCnotifications it is denoted any of the Diameter commands RAR_{IPCAN}, RAR_{SDFD}, RAR_{SPS}, RAR_{UM}.

By $\lambda_{Protocol} = (S_{Protocol}, Act_{Protocol}, \rightarrow_{Protocol}, s_0)$, it is denoted an LTS representing the Diameter QoS resources session where:

$S_{Protocol}$ = { Null, QoSAuthorized}

$Act_{Protocol}$ = {AAR_{IP}, PCCprocedures, PCCnotification, ASR, STR};

$\rightarrow_{Protocol}$ = {Null AAR_{IP} QoS Authorized, QoS Authorized PCCprocedures QoSAuthorized, QoSAuthorized PCCnotifications QoS Authorized, QoSAuthorized STR Null, QoSAuthorized ASR Null }

s_0 = {Null}.

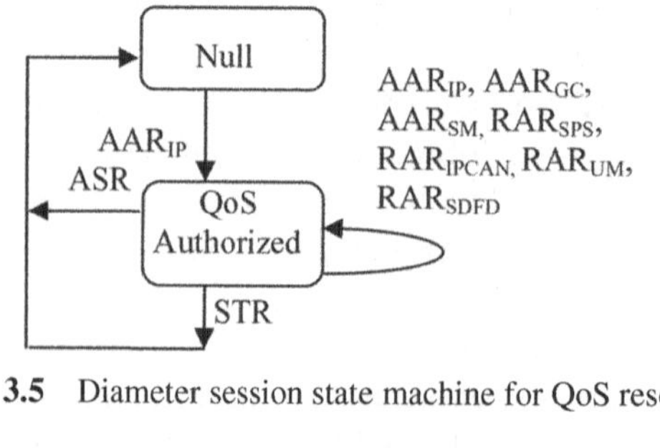

Figure 3.5 Diameter session state machine for QoS resources.

Table 3.2 Bisimulation relation between states

Transitions in T_{AppPCC}	Transitions in $T_{Protocol}$
AppNullreserveSpecificQoSAppQoSAuthorized,	Null AAR_{IP} QoSAuthorized
AppQoSAuthorized PCCrequest	QoSAuthorized PCCprocedures
AppQoSAuthorized	QoSAuthorized
AppQoSAuthorized PCCreports	QoSAuthorized PCCnotifications
AppQoSAuthorized	QoSAuthorized
AppQoSAuthorized allBearersLost	QoSAuthorized ASR Null
NetworkQoSReleased	
AppQoSAuthorized releaseQoSResources	QoSAuthorized STR Null
ApplicationQoSReleased	

The state machines described above need to be synchronized i.e. to expose equivalent behaviour. The concept of weak bisimulation [19] is used to prove that the state machine representing the application view on QoS resources and the state machine representing the Diameter QoS resources session expose equivalent behaviour.

Proposition: The labelled transition systems T_{AppPCC} and $T_{Protocol}$ are weakly bisimilar.

Proof: To prove the bisimulation relation between two Labelled Transition Systems, it has to be proved that there is a bisimulation relation between their states. The states AppNull, NetworkQoSReleased and Application QoSReleased might be absorbed by a single state as no QoS resources are assigned to the user session being in these states.

By U it is denoted a relation between the states of T_{AppPCC} and $T_{Protocol}$ where U = {(AppNull, Null), (AppQoSAuthorized, QoSAuthorized)}. Table 3.2 presents the bisimulation relation between the states of T_{AppPCC} and $T_{Protocol}$. The mapping of "Application-controlled QoS" interface methods onto Diameter commands shows the action's similarity. Based on the bisimulation relation between the states of $T_{AppPCC}$and $T_{Protocol}$ it can be stated that both Labelled Transition Systems expose equivalent behaviour.

Figures 3.6 and 3.7 show examples of procedures for third party QoS management using the APIs of "Application-controlled QoS" and ones for Call Control.

The example shown in Figure 3.6 assumes that a third party is deployed an application that may be used to control the assignment of QoS resources for user sessions.

The application is triggered when the user sends a request for session initiation. It determines the specific QoS that has to be applied to user session (e.g. increased data rates) and invokes the reserveSpecificQoSReq() method

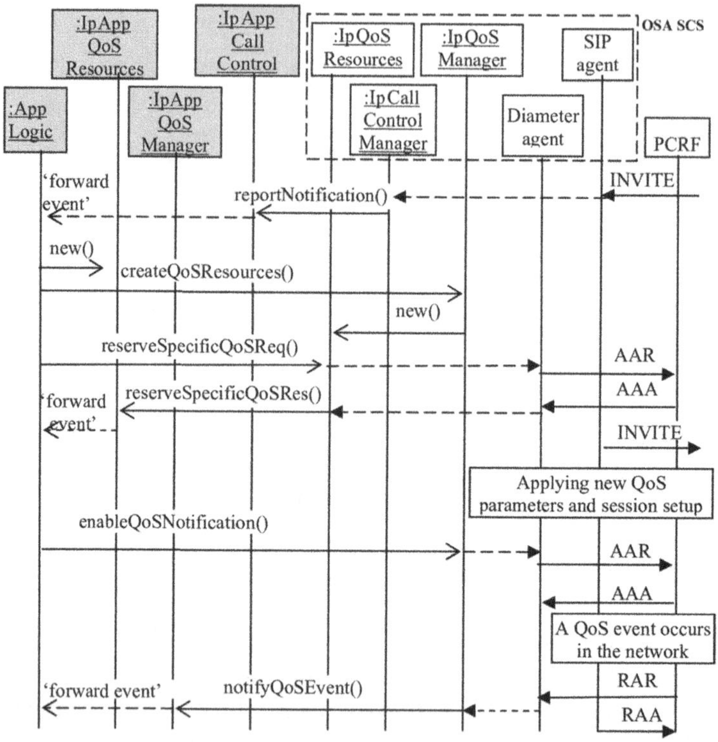

Figure 3.6 Temporary change of QoS features available on user session, subscription for and notification about QoS events.

to apply new QoS features. Assuming that the network supports the requested rates, the user speed is increased to value requested by the application. The application invokes the enableQoSNotifications() method in order to register its interest in receiving notifications about QoS events related on user session. On occurrence of such an event (e.g. the QoS degrades) the application is notified by invocation of notifyQoSEvent() method. The information may be used to apply QoS dependant charging.

Figure 3.7 shows an example of QoS management application that may be used for network resource usage monitoring. The application invokes the setUsageMonitoring() method to set thresholds for monitoring. When a user initiates a session, the network begins monitoring. The application is notified when the thresholds are reached by invocation of the usageMonitoringReport() method. In the example, the application decides to release the QoS resources and the user session.

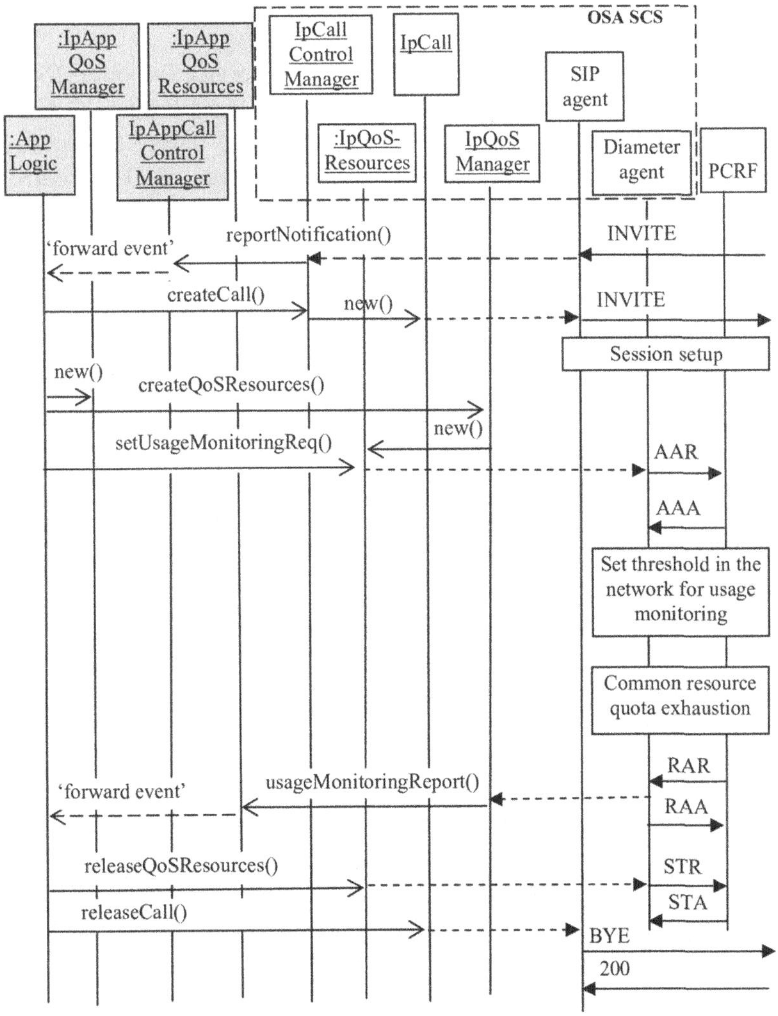

Figure 3.7 Network resources usage monitoring and reporting.

3.5 A Model of Application Server for Open Access to Policy and Charging Control

One of the alternatives for deployment of open access is through Parlay X APIs. In this Section, we present a model of The Parlay X gateway which is a special type of Application Server that makes the conversion between Parlay X APIs and network control protocol.

3.5.1 Abstraction Model

The environment of the Parlay X gateway consists of Service Providers (SPs) that host applications and a node in the network. Using Parlay X APIs, SPs generate requests to be served by the network. The Parlay X gateway communicates with the network node by a control protocol. The control protocol transfers request and answer messages and any notifications from the network.

A typical message exchange template includes application generated requests for initiation and termination of its interest in observing certain conditions in the network. The initiation phase is depicted in Figure 3.8 by the first four messages and the termination phase is consisted of the last four messages. The initiation and termination delineate the notification phase which might be periodic, aperiodic or mixed.

For example, the "Application-driven Quality of Service" (ADQ) is a Parlay X Web service that allows third party applications to dynamically control the QoS available on user connection. During the first phase, the application sets temporary new QoS features to an active user session and subscribes to notifications about QoS events. During the notification phase, any QoS events related to the user session are reported. The application may decide to terminate the user session or to terminate the subscription for QoS events. Another example is the "Terminal location" Web Service

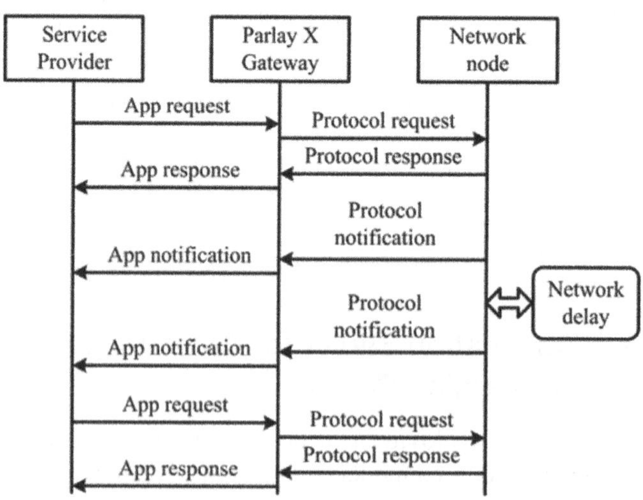

Figure 3.8 Message exchange template.

which allows the application to receive notifications when the user location is changed.

The model in [2] regards "Third Party call" Web Service and respective message exchange initiated by SPs only. For the purposes of our case this is not enough because the incoming traffic is generated by Parlay X applications hosted by SPs, and by the network. The case that is investigated in the chapter is when there are messages of three different classes and respectively three different priorities. These are: messages from applications related to the initial phase (start messages) with lowest priority, notification messages from the network with normal priority and messages from applications related to the third phase (stop messages) with high priority.

A SLA driven architecture is considered. Each SP might include several applications but as the SLAs are agreed between the SPs and the network operator that owns the Parlay X gateway, the number of applications is not important.

The SLA between SP and network operator defines constraints that have to be fulfilled. The constraints include the peak and average number of different application requests and network notifications that should be accepted per time unit, and the maximum delay between application request and response.

To be able to fulfil the restrictions imposed by the network operator and to avoid congestion, the Parlay X gateway implements an Admission Control/Load Balancing mechanism. The Admission Control (AC) is used to protect the Parlay X gateway when there is not enough capacity to process all requests for service. In such case, some of the requests are rejected according for example to the well-known algorithm of Token Bucket (TB). Usually the Parlay X gateway has several processing nodes (converters). In a distributed environment the traffic that has to be served is split between processing nodes equally. There are different algorithms for load balancing (LB).

The Parlay X gateway system is built of *n* ACs and *m* converters as shown in Figure 3.9.

The admission control rejects the nonconforming messages being either SP's requests or network notifications. If a message passes the access control, it is forwarded to appropriately selected converter over the message bus (MB1). The appropriate selection is based on load balancing that distributes uniformly load between converters. Each converter translates application requests into protocol messages, and protocol notifications into application requests. We assume that all converters have same capacity. The Round Robin algorithm stays behind load balancing because of its robustness.

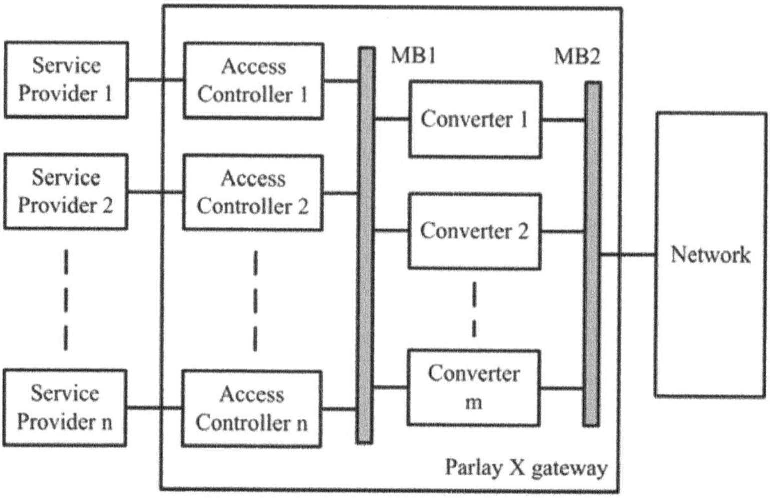

Figure 3.9 Distributed model of Parlay X Gateway structure.

The abstract model of the i-th AC is shown in Figure 3.10, where $V_i(t_k)$ is the number of start and stop messages within the interval $[t_{k-1}, t_k)$. To protect the Parlay X gateway from overload a rough admission control is used (TB1). If a message is rejected, filter F5 passes only rejected stop messages. If a stop message is rejected, the application has to be informed. The filter F1 passes notify messages which are forwarded to the TB2 which prevents the Parlay X gateway from overloading caused by network initiated notifications. The conforming notifications are forwarded to the application. If the accepted message is not a notify message it is filtered by F2 which passes start messages. The start messages are forwarded to the TB3 which controls that the constraint of accepted number of start messages is fulfilled. The accepted start message is forwarded for load balancing.

If an accepted message is neither notification nor start message, it is forwarded to filter F3. In case of stop message filter F4 is used to correlate the stop message with the corresponding start message.

The model of the j^{-th} converter is shown in Figure 3.11 where $Z_j(t_k)$ is the number of all notifications sent by the network within the interval $[t_{k-1}, t_k)$.

Each of the converters is modelled as a single FIFO buffer (Q) with limited size. The message is fetched out the queue, translated by the processing unit (PU) and forwarded either to the network or to the application.

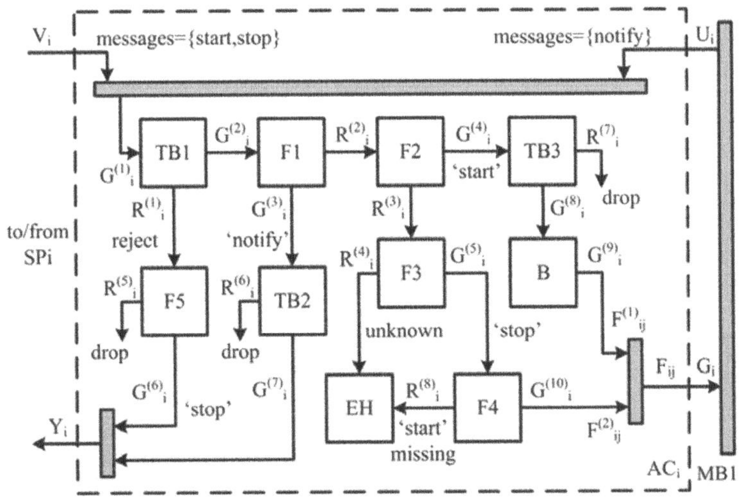

Figure 3.10 Model of i-th access controller.

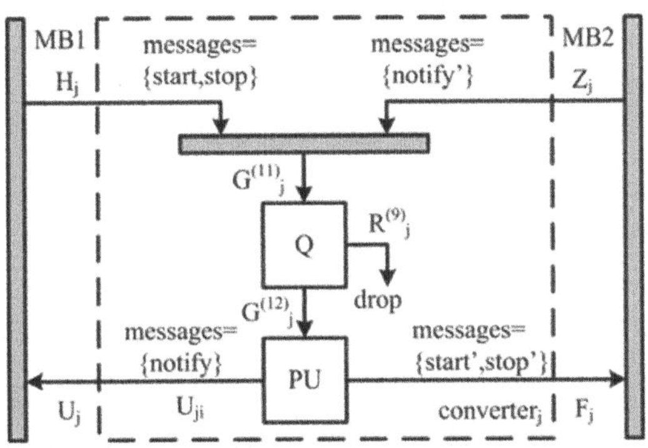

Figure 3.11 Model of j-th converter.

3.5.2 Mathematical Model of Application Server for Open Service Access

This section presents formal descriptions of mechanisms used for admission control, message filtering and load balancing.

The message traffic is observed at regular time intervals $[t_{k-1}, t_k)$. The token bucket model TB(T,ρ,μ) is characterized by volume T, rate ρ and level μ. The equation (3.1) represents the initial token level.

$$\mu(t_0) = T. \tag{3.1}$$

The token level at the end of the time interval $[t_{k-1}, t_k)$ is:

$$\mu(t_k) = \mu(t_{k-1}) + A(t_k) - G(t_k). \tag{3.2}$$

where $A(t_k)$ is the amount of tokens arrived within $[t_{k-1}, t_k)$, and $G(t_k)$ is the amount of tokens given in $[t_{k-1}, t_k)$.

By $N(t_k)$ it is denoted the number of messages arrived to the TB within $[t_{k-1}, t_k)$. Then

$$A(t_k) = \min(T - \mu(t_{k-1}), \rho.(t_k - t_{k-1})). \tag{3.3}$$

$$G(t_k) = \min(\mu(t_{k-1}) + A(t_k), N(t_k)). \tag{3.4}$$

The amount of rejected messages is:

$$R(t_k) = N(t_k) - G(t_k). \tag{3.5}$$

By C = $\{c_i\}$ it is denoted the set of all classes of messages defined, K is the number of message classes, and c_i is the i^{-th} class of messages. Then the classifier function that checks whether given message m belongs to class $c_i$is presented by:

$$d(m, c_j) = i.I(m \in c_j) \tag{3.6}$$

where I is an indicator function i.e. $I(s) = 1$ if s is true, and $I(s) = 0$ otherwise. The number of messages belonging to class $c_j$and passed by a filter is:

$$F(c_j, N(t_k)) = I(N(t_k) > 0).\sum_{n=1}^{N(t_k)} I(j = d(m_n \in c_j)) \tag{3.7}$$

The number of messages rejected by the filter is:

$$R(c_j, N(t_k)) = N(t_k) - F(c_j, N(t_k)) \tag{3.8}$$

The load balancer model LB(σ, M) is characterized by its internal state σ and the set of destinations M that LB is going to distribute the load to. Then the LB initial state is $\sigma(t_0) = random(m)$ where m is the power of set M and $random(m)$ is uniform distribution between 0..m-1.

If $N(t_k) = 0$ then $\sigma(t_k) = \sigma(t_{k-1})$ and the number of messages distributed to the j^{-th} element of M is $F_i(t_k) = 0$. If $N(t_k) > 0$ then the index of the element of M that is going to be the destination for the n^{-th} message within $[t_{k-1}, t_k)$ is

$$b_n = 1 + (\sigma(t_{k-1}) + n) \bmod m, \quad n = 1..N(t_k) \tag{3.9}$$

The number of messages addressed to the j^{-th} destination within the period $[t_{k-1}, t_k)$ becomes:

$$F_j(t_k) = \sum_{n=1}^{N(t_k)} I(b_n = j) \tag{3.10}$$

Combining the above cases $N(t_k) = 0$ and $N(t_k) > 0$ gives the total number of messages balanced toward j^{-th} destination

$$F_j(t_k) = I(N(t_k) > 0). \sum_{n=1}^{N(t_k)} I(b_n = j) \tag{3.11}$$

Let any message of class c_j require $w^{(j)}$ amount of service provided by the converter it is dispatched to. Then the state of the buffer can be recurrently presented as

$$q(t_k) = q(t_{k-1}) - G(t_k) + A(t_k) \tag{3.12}$$

where $G(t_k)$ is the amount of messages fetched out of the buffer and $A(t_k)$ is the amount of the placed ones and $q(t_0) = 0$. Having ζ of the total converter's capacity dedicated to message conversion and a hard limit set on the length of the buffer, denoted by Q_{max}, it is easy to define for $G(t_k)$ and $A(t_k)$ the following

$$\begin{aligned} G(t_k) &= I\left(q(t_{k-1}) + N(t_k) > 0\right) \\ &\times \max_m \left(\zeta \cdot (t_k - t_{k-1}) \geq \sum_{l=1}^{m} \sum_{j=1}^{K} w^{(j)} \cdot I\left(q[l] \in c_j\right) \right) \end{aligned} \tag{3.13}$$

$$A(t_k) = \min\left(N(t_k), G(t_k) + Q_{\max} - q(t_{k-1})\right) \tag{3.14}$$

It is trivial to observe that the amount of losses caused by finite buffer in the converter within $[t_{k-1}, t_k)$ is

$$R(t_k) = N(t_k) - A(t_k) \tag{3.15}$$

By c_1 it is denoted the class of start messages, by c_2 the class of stop messages and by c_3 the class of notify messages. Let S denote the set of active subscriptions. The function $sid(m_n)$ extracts out of the n^{-th} stop message m within the interval having identifier of the subscription to be stopped. The function $a(sid(m_n))$ extracts the index of converter which is engaged with given subscription. Then the message flows in the i^{-th} access controller shown in Figure 3.10 are described by the equations from (3.16) to (3.38).

$$G_i^{(1)}\left(t_k\right) = U_i\left(t_k\right) + V_i\left(t_k\right). \tag{3.16}$$

$$G_i^{(2)}(t_k) = \min\left(\mu_i^{(1)}(t_{k-1}) + \min\left(\rho_i^{(1)} \cdot (t_k - t_{k-1}),\right.\right.$$
$$\left.\left.T_i^{(1)} - \mu_i^{(1)}(t_{k-1})\right), G_i^{(1)}(t_k)\right). \tag{3.17}$$

$$R_i^{(1)}\left(t_k\right) = G_i^{(1)}\left(t_k\right) - G_i^{(2)}\left(t_k\right). \tag{3.18}$$

$$G_i^{(3)}\left(t_k\right) = I\left(G_i^{(2)}\left(t_k\right) > 0\right) \cdot \sum_{n\,=\,1}^{G_i^{(2)}(t_k)} I\left(3 = d\left(m_n \in c_3\right)\right). \tag{3.19}$$

$$R_i^{(2)}\left(t_k\right) = G_i^{(2)}\left(t_k\right) - G_i^{(3)}\left(t_k\right). \tag{3.20}$$

$$G_i^{(4)}\left(t_k\right) = I\left(R_i^{(2)}\left(t_k\right) > 0\right) \cdot \sum_{n\,=\,1}^{R_i^{(2)}(t_k)} I\left(1 = d\left(m_n \in c_1\right)\right). \tag{3.21}$$

$$R_i^{(3)}\left(t_k\right) = R_i^{(2)}\left(t_k\right) - G_i^{(4)}\left(t_k\right). \tag{3.22}$$

$$G_i^{(5)}\left(t_k\right) = I\left(R_i^{(3)}\left(t_k\right) > 0\right) \cdot \sum_{n\,=\,1}^{R_i^{(3)}(t_k)} I\left(2 = d\left(m_n \in c_2\right)\right). \tag{3.23}$$

$$R_i^{(4)}\left(t_k\right) = R_i^{(3)}\left(t_k\right) - G_i^{(5)}\left(t_k\right). \tag{3.24}$$

$$G_i^{(6)}\left(t_k\right) = I\left(R_i^{(1)}\left(t_k\right) > 0\right) \cdot \sum_{n\,=\,1}^{R_i^{(1)}(t_k)} I\left(2 = d\left(m_n \in c_2\right)\right). \tag{3.25}$$

$$R_i^{(5)}\left(t_k\right) = R_i^{(1)}\left(t_k\right) - G_i^{(6)}\left(t_k\right). \tag{3.26}$$

$$G_i^{(7)}(t_k) = \min\left(\mu_i^{(2)}\left(t_{k-1}\right) + \min\left(\rho_i^{(2)} \cdot \left(t_k - t_{k-1}\right),\right.\right.$$
$$\left.\left.T_i^{(2)} - \mu_i^{(2)}\left(t_{k-1}\right)\right), G_i^{(3)}\left(t_k\right)\right). \tag{3.27}$$

$$R_i^{(6)}\left(t_k\right) = G_i^{(3)}\left(t_k\right) - G_i^{(7)}\left(t_k\right). \tag{3.28}$$

$$G_i^{(8)}(t_k) = \min\left(\mu_i^{(3)}(t_{k-1}) + \min\left(\rho_i^{(3)} \cdot (t_k - t_{k-1}),\right.\right.$$
$$\left.\left. T_i^{(3)} - \mu_i^{(3)}(t_{k-1})\right), G_i^{(4)}(t_k)\right). \quad (3.29)$$

$$R_i^{(7)}(t_k) = G_i^{(4)}(t_k) - G_i^{(8)}(t_k). \quad (3.30)$$

$$F_{ij}^{(1)}(t_k) = I\left(G_i^{(8)}(t_k) > 0\right) \cdot \sum_{n=1}^{G_i^{(8)}(t_k)} I\left(j = 1 + \left(\sigma_i(t_{k-1}) + n\right) \mathrm{modM}\right). \quad (3.31)$$

$$G_i^{(9)}(t_k) = \sum_{j=1}^{M} F_{ij}^{(1)}(t_k). \quad (3.32)$$

$$G_i^{(10)}(t_k) = I\left(G_i^{(5)}(t_k) > 0\right) \cdot \sum_{n=1}^{G_i^{(5)}(t_k)} I\left(sid(m_n) \in S\right). \quad (3.33)$$

$$R_i^{(8)}(t_k) = G_i^{(5)}(t_k) - G_i^{(10)}(t_k). \quad (3.34)$$

$$F_{ij}^{(2)}(t_k) = I\left(G_i^{(5)}(t_k) > 0\right) \cdot \sum_{n=1}^{G_i^{(5)}(t_k)} I\left(sid(m_n) \in S\right) \text{ x}$$
$$\text{x} I\left(j = a\left(sid(m_n)\right)\right). \quad (3.35)$$

$$F_{ij}(t_k) = F_{ij}^{(1)}(t_k) + F_{ij}^{(2)}(t_k). \quad (3.36)$$

$$R_i(t_k) = \sum_{j=2}^{8} R_i^{(j)}(t_k). \quad (3.37)$$

$$Y_i(t_k) = G_i^{(6)}(t_k) + G_i^{(7)}(t_k). \quad (3.38)$$

The amount of messages dispatched by the j^{-th} converter to the i^{-th} access controller is denoted by U_{ji}. The message flows in the j^{-th}converter are described by the equations from (3.39) to (3.45).

$$H_j(t_k) = \sum_{i=1}^{N} F_{ij}(t_k). \quad (3.39)$$

$$G_j^{(11)}(t_k) = H_j(t_k) + Z_j(t_k). \tag{3.40}$$

$$G_j^{(12)}(t_k) = I\left(q_j(t_{k-1}) + G_j^{(11)}(t_k) > 0\right)$$
$$\times \max_m \left(\kappa_j \cdot (t_k - t_{k-1}) \geq \sum_{l=1}^{m} \sum_{n=1}^{K} w^{(n)} \cdot I(q_j[l] \in c_n)\right). \tag{3.41}$$

$$R_j^{(9)}(t_k) = G_j^{(11)}(t_k) -$$
$$- \min\left(G_j^{(11)}(t_k), Q_{\max j} + G_j^{(12)}(t_k) - q_j(t_{k-1})\right). \tag{3.42}$$

$$U_j(t_k) = I\left(G_j^{(12)}(t_k) > 0\right)$$
$$\times \sum_{n=1}^{G_j^{(12)}(t_k)} I(3 = d(m_n \in c_3)) \cdot I(sid(m_n) \in S). \tag{3.43}$$

$$U_{ji}(t_k) = I\left(U_j(t_k) > 0\right) \cdot \sum_{n=1}^{U_j(t_k)} I(i = a(sid(m_n))). \tag{3.44}$$

$$F_j(t_k) = G_j^{(12)}(t_k) - U_j(t_k). \tag{3.45}$$

So the message flow in U_i shown in Figure 3.10 which represents the number of messages sent by all the converters to the i^{-th} access controller becomes

$$U_i(t_k) = \sum_{j=1}^{M} U_{ji}(t_k) \tag{3.46}$$

The following notations are used:

- N – number of SPs, respectively AC
- M – number of converters
- κ_i- capacity of i^{-th} access controller dedicated to access control function
- κ_j- capacity of j^{-th}converter dedicated to message conversion.

The gateway utilization and throughput in $[t_{k-1}, t_k)$ are given by equations (3.47) and (3.48) respectively.

$$\eta(t_k) = \frac{\sum_{i=1}^{N} Y_i(t_k) + \sum_{j=1}^{M} F_j(t_k)}{(t_k - t_{k-1}).\left(\sum_{j=1}^{M} \kappa_j + \sum_{i=1}^{N} \kappa_i\right)}. \tag{3.47}$$

$$\gamma(t_k) = \eta(t_k) \cdot \left(\sum_{j=1}^{M} \kappa_j + \sum_{i\,=\,1}^{N} \kappa_i \right). \tag{3.48}$$

The loss function for the Parlay X gateway is given by:

$$L(t_k) = \sum_{i\,=\,1}^{N} (\alpha_i(t_k) \sum_{p=4}^{8} R_i^{(p)}.(t_k)) + \sum_{j=1}^{M} \beta_j(t_k).R_j^9(t_k)). \tag{3.49}$$

The parameters $\{\alpha_i, \beta_j, i = 1...N, j = 1...M\}$ are nonnegative functions of time assigning relative weights given to various losses. The first parameter imposes a penalty on lost traffic at the admission control for the i^{-th} TB and the second parameter imposes a penalty on losses in the j^{-th} converter. The problem is to find a control policy that minimizes this function.

By $\rho_{gi}^{(2)}$ it is denoted the guaranteed rate of notification from the SLA between i^{-th} SP and the network operator and as it is fixed it doesn't vary in time. The focus is set on the losses of notifications as far as the converters had spent part of their capacity to perform conversion. Thus the control policy is on the actual rate of tokens for the TB2s as follow:

$$\rho_i^{(2)}(t_k) = \rho_{gi}^{(2)}(t_k) + \left\lfloor \max(\kappa_i(t_k - t_{k-1}) - G_i^{(1)}(t_k), 0).\frac{U_i(t_k)}{V_i(t_k) + U_i(t_k)} \right\rfloor \tag{3.50}$$

The left side of equation (3.50) presents the rate of tokens that will be granted to i^{-th} SP for the interval $[t_k, t_{k+1})$ especially for messages of notification type. The first component on the right side is the guaranteed rate of tokens from SLA, and the second component represents the part of available resources of i^{-th} access controller that are proportionally engaged with the notifications flow of messages.

3.5.3 Simulation Parameters

The simulation is done on simplified Parlay X model with three classes of messages and four converters. The parameters used are provided by a mobile operator. The capacity of the gateway is 800 requests per second which is equally distributed between the converters. The behaviour of each SP is modelled by Markov Modulated Poisson Process (MMPP). New application requests are generated according to four-state MMPP. Changes between different states are uniformly distributed and occurred according to Poisson process with mean 4s. The time intervals between start messages generated

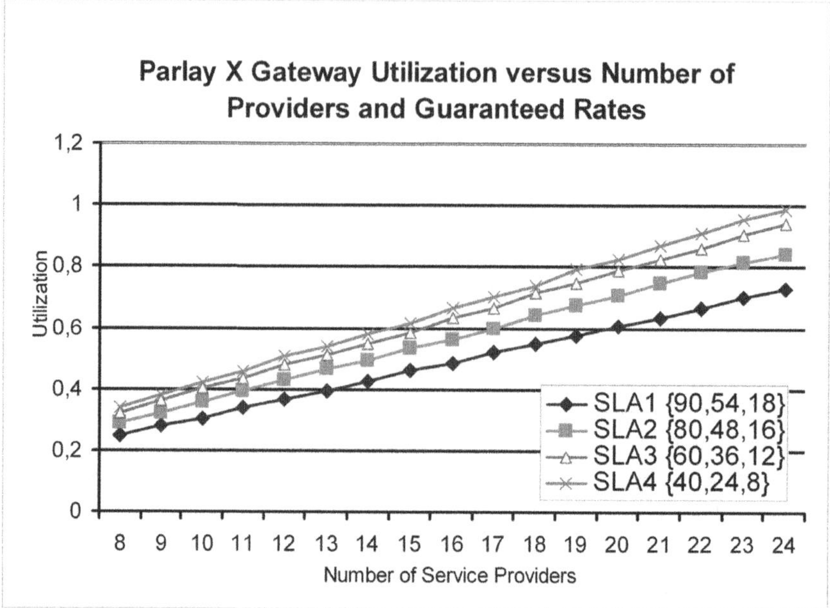

Figure 3.12 Model of j-th converter.

by each SP are exponentially distributed as the arrival process in the context of web services [16] with mean 150 seconds. During an application session the time intervals between notify messages generated by the network are exponentially distributed with mean 50 *s*. The SPs traffic is policed by access controllers whose conforming outputs are multiplexed between 4 converters. The token rate is equal to the guaranteed rate and the bucket size is determined by the peak rate. Initially, $\mu_i(t_0) = T_0$, $q_i(t_0) = 0$. The length of the interval for observation $[t_k$-$t_{k-1})$ is set to 100 *ms*. The processing time for a single request/ response in a converter is 5 *ms*.

The aim of simulation is to evaluate the Parlay X gateway utilization setting different values of guaranteed rates and fixed peak rates. The guaranteed rates for a given SP define the constraints for preventing the Parlay X gateway from overloading (GR_1), for the rate of notify messages (GR_2), and for the rate of start messages (GR_3). The processing capacity of the Parlay X Gateway must be distributed between different types of messages, where the overall message peak rate (PR_1) must be spread between notification messages (PR_2), start messages (PR_3), and stop messages.

3.5.4 Results and Discussion

The simulation is run in a space of Service Level Agreements (SLAs) where each one is consisted of tuple (PR_i, GR_i) for every TB_i of given access controller. Figure 3.12 summarizes the outcome of the simulation. The Parlay X gateway utilization is evaluated as a function of the number of SPs and guaranteed rates. The guaranteed rates of messages per second as assigned by the mobile operator are included into four different SLAs. The values of peak rates PR_1=100, PR_2=60, PR_3=20 are limited by the converter capacity.

The simulation results show that the utilization depends both on the number of SPs and on the specific values of rates in SLAs. In case the network operator sets the congestion threshold value of 80%, it is most likely that the appropriate choice is to have 22 SPs applying second type or third type of SLA. Applying adaptive control leads to higher utilization than the case without control. The average throughput gain is about 8%.

3.6 Conclusion

The open access to resource management functions enforces development of new attractive customized applications. The usage of APIs instead of telecommunication protocols makes the process of service creation easier and closed to application development in internet.

The standardized Policy and Charging Control architecture provides capabilities for effective resource management. As this architecture is expanding continuously with every new Release, the standardized capabilities for open access to QoS control do not support most of the PCC functions.

The suggested approach to design of APIs for open access to resource management is based on the requirements to open access to PCC functions. The mapping of APIs onto Diameter protocol illustrates their applicability in IMS. The functional verification of application server for open access to resource management is proved by formal description of its functional behaviour and usage of the concept of weak bisimulation. The application server model is used to evaluate its performance. The proposed an adaptive mechanism for load control increases the application server utilization.

The open access to QoS control provides more flexibility in resource management as far as the QoS provisioning is one of the main requirements to the all IP based networks. The benefits for the end users are more advanced applications. Exploiting an open access technology is opportunity for third party service providers to gain a telecom services market share. Opening the network allows the network provider to increase its revenues by diversification of its services portfolio.

References

[1] Akhatar, H. 2009. "System and Method for Providing Quality of Service Enablers for Third Party Applications", Patent application number: 20090154397.

[2] Anderson, J., M. Kihl, D. Sobirk, 2004. "Overload Control of a Parlay X Application Server", *Proc. of SPECTS'04*, pp.821–828.

[3] Bartolini, N., G. Bongiovanni, S. Silvestri, 2009. "Self-through self-learning: Overload control for distributed web systems", Computer Networks, 53, pp.727–743.

[4] Bormann, F., Braun, A., Flake, S., Tacken, J. 2009. "Towards a Policy and Charging Control Architecture for Online Charging", Proc. of ICAINA, 2009, pp. 524–530.

[5] Brebner, P.; L. O'Brien, J. Gray, 2009, "Performance modelling power consumption and carbon emissions for Server Virtualization of Service Oriented Architectures (SOAs)", Proc. of EDOCW'2009, pp.92–99.

[6] Chappell, D., D. Berry, 2007, "SOA - Ready for Primetime: The Next-Generation, Grid-Enabled Service-Oriented Architecture", SOA Magazine, Issue X, http://www.soamag.com/I10/0907–1.php.

[7] Corici M., Magedanz T., Vingarzan D., Pampu C., Zhou Q. (2010) "Access Network Reselection based on Momentary Resources in a Converged Wireless Environment", Next Generation Networking Symposium'2010.

[8] Darvishan, A., H. Yeganeh, K. Bamasian, H. Ahmadian, 2010. "OSA Parlay X gateway architecture for third party operators participation in next generation networks", Proc. of ICACT'10, pp. 75–80.

[9] De, P., P. Chodhury, S. Choudhury, 2010. "A Framework for Performance Analysis of Client/Server Based SOA and P2P SOA", Proc. of ICCNT'2010, pp.79–83.

[10] Gouveia, F., Wahle S., Blum, N. & Megedanz, T. (2009). Cloud computing and EPC / IMS integration: new value-added services on demand, IMMCC' 2009.

[11] Good, R., Gouveia, F., Ventura, N. & Magedanz, T. 2010. "Session-based end-to-end policy control in 3GPP evolved packet system", International Journal of Communication Systems, Special Issue: Part 1: Next Generation Networks, vol. 23, issue 6–7, pp. 861–883.

[12] Good, R. & Ventura, N. 2009. "Application-driven Policy-Based Resource Management for IP multimedia subsystems", International Conference on Testbeds and Research Infrastructures for the Development of Networks & Communities, TridentCom'2009, pp.1–9.

[13] Jain, M. & Prokopi, M. 2008. "The IMS 2.0 Service Architecture", Proc. of NGMAST '08, pp.3–9.

[14] Kassev, K., Y. Mihov, A. Kalaydzhieva, B. Tsankov, 2010. "CAC Dimensioning for VoIP Trafficover Wireless Access Networks: From Network to Application-specific perspective," International Journal on Advances in Networks and Services, ISSN: 1942–2644, vol. 3, no. 3 & 4, 2010, pp. 333–345.

[15] Mathur, V., S. Dhopeshwarkar, V. Apte, 2009. "MASTH Proxy: An Extensible Platform for Web Overload Control", Retrieved from: http://www2009.org/ proceedings/pdf/p1113.pdf.

[16] Muscariello, M., M. Mellia, M. Meo, M. Marsan, R. Lo Cigno, 2005. "Markov models of internet traffic and a new hierarchical MMPP model", Computer Communications, vol. 28 , issue 16, pp. 1835–1851.

[17] Musthaq, S., Salem, O., Lohr, C. & Gravey, A. 2008. "Policy-Based QoS Management for Multimedia Communication", Retrieved from https://portail.telecom-bretagne.eu/publi/public/ download.jsp?id...542.6.

[18] Ouellette, S., Marchand, L. & Pierre, S. 2011. "A potential evolution of the policy and charging control/QoS architecture for the 3GPP IETF-based evolved packet core", IEEE Communications Magazine, vol. 49, issue 5, pp. 231–239.

[19] Panangaden P., 2009. "Notes on Labelled Transition Systems and Bisimulation", Retrieved from http://www.csmcgill.ca/~prakash/Courses /comp330/Notes/lts09.pdf.

[20] Riad, A., A. Hassan, Q. Hassan, 2010. "Design of SOA-based Grid Computing with Enterprise Service Bus", International Journal on Advances in Information Sciences and Service Sciences, vol. 2, No 1, pp.71–82.

[21] Stojanovic, M., Rakas, S. and Acimovic-Raspopovic, V., 2010. "End-to-end quality of service specification and mapping: The third party approach", Computer Communications, vol. 1, pp. 1354–1368.

[22] Wang, Y., Liu, W. & Guo, W., 2010. "Architecture of IMS over WiMAX PCC and the QoS mechanism", Proc. of ICWMNN'10, pp.159–162.

[23] Yang, J. & Park, H., 2008. "A Design of Open Service Access Gateway for Converged Web Service", Proc. of ICCT'2008, pp.1807–1810.

[24] Yordanov V., G. Iliev, V. Poulkov, 2010. "Application of beamforming in wireless local area computer networks", International Journal on Information Technologies and Security, No 2, 2010, pp. 3–16.
[25] Zhang Qi-zhi, 2008. "On overload control of parlay application server in next generation network", Journal of China Universities of Posts and Telecommunications, vol. 15, issue 1, pp. 43–47.

4

Call-Level Performance Evaluation of Cognitive and AMC-Enabled Wireless Access Networks

Yakim Mihov, Kiril Kassev and Boris Tsankov
Department of Communication Networks; Technical University of Sofia, Bulgaria
E-mail: Kiril Kassev <kmk@tu-sofia.bg>

4.1 Introduction

Radio spectrum is a limited and precious resource in wireless access networks. Since the demand for spectrum resources grows increasingly because of newly emerging broadband wireless services, efficient spectrum utilization becomes a matter of great importance. There are many different methods and approaches for increasing the spectrum utilization efficiency, such as hierarchical structures with microcells and femtocells, multi-hop connections, new multiple access methods and scheduling algorithms, adaptive modulation and coding (AMC), multiple-input multiple-output (MIMO) antenna systems, inter-cell interference management, handover management, dynamic spectrum access (DSA) and cognitive radio (CR) networks in particular, etc. This chapter focuses mainly on DSA and AMC-enabled wireless access networks with respect to the efficient utilization of the transmission resources and the quality-of-service (QoS) provisioning in these networks.

The chapter's first half is devoted to CR networks and the chapter's second half is devoted to AMC-enabled networks. In both cases, call-level (network-level) performance evaluation is considered.

Vladimir Poulkov and Ramjee Prasad (Eds.), Resource Management in Future Internet, 75–110.

4.2 QoS Provisioning and Performance Analysis of Cognitive Radio Networks

The CR concept proposes to push efficiency in spectrum access and resource allocation beyond its traditional limits, by introducing spectrum sharing, coexistence, and cooperation among heterogeneous wireless networks. This approach has been worldwide recognized by standardization and regulation bodies, and a number of wireless technologies could benefit from it. Open research issues cover a wide range of system aspects, from the hardware component up to the network layer design. Technical, economical, and regulatory challenges need to be addressed as well.

4.2.1 Overview of DSA

DSA is a new paradigm for spectrum regulation which is intended to improve the overall spectrum utilization and thus to alleviate the artificially created scarcity of spectrum resources caused by the traditional static command-and-control approach for spectrum regulation. DSA strategies can be broadly categorized under three essential models: *dynamic exclusive use model* with two approaches – spectrum property rights and dynamic spectrum allocation; *open sharing model* (spectrum commons model); *hierarchical access model* with two approaches – hierarchical spectrum overlay and hierarchical spectrum underlay [1].

4.2.1.1 Dynamic exclusive use model

This model maintains the basic structure of the current spectrum regulation policy, i.e., spectrum bands are licensed to services for exclusive use. The main idea is to introduce flexibility to improve spectrum efficiency. Two approaches have been proposed under this model: spectrum property rights and dynamic spectrum allocation. The former approach allows licensees to sell and trade spectrum and freely to choose technology. Economy and market will thus play a more important role in driving towards the most profitable use of this limited resource. It should be noted that even though licensees have the right to lease or share the spectrum for profit, such sharing is not mandated by the regulation policy. The latter approach, dynamic spectrum allocation, was brought forth by the European DRiVE project. It aims to improve spectrum efficiency through dynamic spectrum assignment by exploiting the spatial and temporal traffic statistics of different services. In other words, in a given region and at a given time, spectrum is allocated to services for exclusive use. This allocation, however, varies at a much faster scale than the current policy.

Based on an exclusive-use model, these approaches cannot eliminate white space in spectrum resulting from the bursty nature of wireless traffic.

4.2.1.2 Open sharing model

Also referred to as spectrum commons, this model employs open sharing among peer users as the basis for managing a spectral region. Advocates of this model draw support from the phenomenal success of wireless services operating in the unlicensed industrial, scientific, and medical (ISM) radio band (e.g., Wi-Fi). Centralized and distributed spectrum sharing strategies have been initially investigated to address technological challenges under this spectrum management model.

4.2.1.3 Hierarchical access model

This model adopts a hierarchical access structure with primary users (PUs) and secondary users (SUs). The basic idea is to open licensed spectrum to SUs while limiting the interference perceived by PUs (licensees). Two approaches to spectrum sharing between PUs and SUs have been proposed: hierarchical spectrum underlay and hierarchical spectrum overlay.

The hierarchical spectrum underlay approach imposes severe constraints on the transmission power of the SUs so that they operate below the noise floor of the PUs. By spreading transmitted signals over a wide frequency band (e.g., ultra-wideband (UWB) or spread-spectrum techniques), SUs can potentially achieve a short-range high data rate with extremely low transmission power. Based on a worst-case assumption that PUs transmit all the time, this approach does not rely on detection and exploitation of spectrum white space.

The hierarchical spectrum overlay approach was first envisioned by Joseph Mitola III (see [2], [3]) under the term spectrum pooling and then investigated by the DARPA Next Generation (XG) program under the term opportunistic spectrum access (OSA). Differing from the hierarchical spectrum underlay, this approach does not necessarily impose severe restrictions on the transmission power of the SUs, but rather on when and where they may transmit. It directly targets at spatial and temporal spectrum white space by allowing SUs to identify and exploit local and instantaneous spectrum availability in a nonintrusive manner. The basic components of OSA include spectrum opportunity identification, spectrum opportunity exploitation, and regulatory policy. Opportunity identification involves accurately identifying and intelligently tracking idle frequency bands that are dynamic in both time and space. Opportunity exploitation involves transmitting on the identified white spaces. Regulatory policy defines the basic etiquette for SUs to ensure compatibility with legacy systems. The overall design objective of OSA is to provide sufficient benefit to SUs while protecting spectrum licensees from interference. The tension between the SUs' desire for performance and the PUs' need for protection dictates the interaction across opportunity identification, opportunity exploitation, and regulatory policy.

Compared to the dynamic exclusive use and open sharing models, the hierarchical access model is perhaps the most compatible with current spectrum management policies and legacy wireless systems. Furthermore, the spectrum underlay and overlay approaches can be employed simultaneously to further improve spectrum efficiency.

A detailed and elaborate taxonomy of various possible approaches and models of DSA is presented in [4]. The interested reader can find definitions and explanations of key concepts in the fields of spectrum management (and DSA), CR, policy-defined radio, adaptive radio, software-defined radio, and related technologies in [5].

4.2.2 Overview of Basic Spectrum Management Functions in CR Networks

CR is the key enabling technology for DSA. It is a type of radio in which communication systems are aware of their environment and internal state and can make decisions about their radio operating behavior based on that information and predefined objectives [5]. CR uses software-defined radio (SDR), adaptive radio, and other technologies to adjust automatically its behavior or operations to achieve desired objectives. CR concepts can be applied to a variety of wireless communications scenarios [6, Chapter 1], e.g., next generation wireless networks, coexistence of different wireless technologies, eHealth services, intelligent transportation systems, emergency networks, military networks, etc. One of the most prominent applications of CR is in DSA networks. A comprehensive survey of the state-of-the-art in cognitive radio technology and its application to DSA is presented in [7]. Recent developments and open research issues related to spectrum management in CR networks are discussed in [8]. While CR networks are not limited only to DSA, as already stated above, we will nonetheless focus on CR networks for DSA in compliance with the hierarchical spectrum overlay approach.

CR networks are envisioned to provide high bandwidth to mobile users via heterogeneous wireless architectures and dynamic spectrum access techniques. This goal can be achieved only through dynamic and efficient spectrum management techniques. CR networks impose unique challenges because of the high fluctuation in the available spectrum and the diverse QoS requirements of various applications. In order to address these challenges, the CR network must be able to determine which segments of the spectrum are available, to select the most favorable idle channels for transmission, to coordinate spectrum access among the cognitive users, and to vacate channels on which licensed users begin to transmit in order not to violate the interference constraints imposed by the licensed network. These capabilities can be realized through spectrum management functions that address five main tasks: spectrum sensing, spectrum analysis, spectrum decision, spectrum sharing, and spectrum mobility (aka spectrum handover or spectrum handoff).

4.2.2.1 Spectrum Sensing

Spectrum sensing in very broad terms involves the detection (by a given receiver) of the presence of a transmitted signal of interest [9, Chapter 4]. The ability of the CR network to access white spaces that appear dynamically is predicated upon its ability to detect these white spaces in the first place. Moreover, while SUs are occupying white spaces, they must be on the lookout for the return of PUs, i.e., continuous monitoring of the spectrum may be necessary. There are some essential design requirements on any spectrum sensing technique. The *first objective* is to detect accurately the presence of existing transmitters - in this case the PUs. The literature in the field tends to express the problem of postulating whether a PU is present or not as a hypothesis test. The null hypothesis states that there is no signal in a certain spectrum band, i.e., there is just noise. The alternative hypothesis states that there is more than noise there and the signal includes transmission from licensed users. The problem is therefore all about determining whether the null or the alternative hypothesis is true, i.e., to detect the presence of the PU. If there is no PU signal on a sensed channel and the CR observes that there is signal, this is known as a false alarm. If there is PU signal on a sensed channel and the CR does not observe its presence, this is known as a missed detection. The *aim* of any spectrum sensing system is to make sure that the number of false alarms and the number of missed detections is as low as they possibly can be and hence the detection rate is as high as possible. A false alarm can lead to a missed transmission opportunity. While this is undesirable, a missed detection has even more serious consequences, since it could cause intolerable interference to the undetected PU. The *second objective* of any spectrum sensing technique is to be valid and usable over the appropriate detection range. To a certain extent, this is an extension of the accuracy requirement expressed in terms of the range from a primary transmitter at which the PU signal must be detected. SUs must be able to detect the presence of a primary transmitter over its decodability range. In other words, the spectrum sensing system cannot be less sensitive than the PU receivers. The *third objective* of any sensing technique is to sense the presence of PU signals in a timely fashion. If the observation process is long, then this can lead to very inefficient use of the available white spaces. In some cases, the opportunity to use a white space may have passed by the time it has been discovered. Furthermore, if the time needed for SUs to observe that a PU has begun to transmit is very large, the amount of interference incurred by the primary network would be unacceptable. Therefore, the challenge is not just about accurately determining if spectrum is free, but doing it in a timely manner. The *fourth objective* of any spectrum sensing technique is to be resilient to interference and noise. The objectives of accuracy, sensitivity, and timeliness may be particularly difficult to meet in the presence of interference and/or noise.

Some well-known spectrum sensing techniques (i.e., energy detection, matched filter, cyclostationary detection, and energy-based wavelet detection) are discussed and profoundly analyzed in [10, Chapter 7]. The *energy detection* method is optimal for detecting any unknown zero-mean constellation signals. In the energy detection approach, the radio frequency energy or the received signal strength indicator (RSSI) is measured over an observation time to determine whether the spectrum is occupied or not. Although this approach can be implemented without any a priori knowledge of the PU signal, it has some drawbacks. A PU signal can only be detected if the detected energy is above the threshold. The threshold selection for energy detection could also be problematic since it is highly susceptible to the changing background noise and especially interference level. Another challenging issue is that the energy approach cannot distinguish PUs from other SUs sharing the same channel. A *matched filter* is an optimal detection method as it maximizes the signal-to-noise ratio (SNR) of the received signal in the presence of additive Gaussian noise. However, a matched filter requires a priori knowledge of the PU signals at both physical and medium access control (MAC) layers, which may not always be readily available to the CR network. Consequently, the matched filter mechanism is a sort of feature-based spectrum sensing. A matched filter is facilitated by correlating a known signal with an unknown signal to detect the presence. The use of the matched filters predicates the need for a dedicated receiver for every kind of primary system, which increases complexity. *Cyclostationary* detection exploits the built-in periodicity of the modulated PU signals, e.g., sine wave carriers, pulse trains, repeating spreading, hopping sequences, cyclic prefixes, etc. The main advantage of the cyclostationary detection is that it can distinguish the noise energy from the signal energy. Thus, a cyclostationary detector is more robust to noise uncertainty than an energy detector. Moreover, it can work with lower SNR than energy detectors. However, the implementation of cyclostationary detection is more complicated than the implementation of energy detection and the cyclostationary detectors need longer observation time than the energy detectors, which means that some spectrum holes with short time duration may not be exploited efficiently. The *wavelet detection* is effective for detecting wideband signals. It offers advantages in terms of both implementation cost and flexibility in adapting to dynamic spectrum as opposed to the conventional use of multiple narrowband bandpass filters. In order to identify the locations of vacant frequency bands, the entire wideband is modeled as a train of consecutive frequency sub-bands where the power spectral characteristic is smooth within each sub-band but changes abruptly on the border of two neighboring sub-bands. By employing a wavelet transform of the power spectral density of the observed signal, the singularities of the power spectral density can be located and thus the vacant frequency bands can be found. A critical challenge of implementing the wavelet approach in practice is the high sampling rates for characterizing large bandwidth. If multiple systems coexist in spectrum, it is difficult for the wavelet detector to identify these multiple systems because of the inter-system interference environment.

The spectrum sensing techniques described above are based on *primary transmitter detection.* Other possible techniques for spectrum sensing are spectrum sensing based on *primary receiver detection* and spectrum sensing based on *interference temperature management* [11, Chapter 1]. The former approach, primary receiver detection, may be feasible for TV receivers only, or further hardware, such as a supporting sensor network, in the area with the primary receivers is required. The objective of this approach is to detect the PUs that are receiving data within the communication range of a cognitive user. The primary receiver usually emits local oscillator leakage power from its RF front-end when it receives signals from a primary transmitter. In order to determine the spectrum availability, the primary receiver detection method exploits this local oscillator leakage power instead of the signal from the primary transmitter, and detects the presence of the primary receiver directly. The latter approach, interference temperature management, is based on a new model for measuring interference, referred to as interference temperature. The interference temperature limit is the amount of new interference that the primary receiver could tolerate. As long as cognitive users do not exceed this limit, they can use the spectrum band. Although this model is best fitted for the objective of spectrum sensing, the difficulty lies in accurately determining the interference temperature limit for each location-specific case. There is no practical way for a cognitive user to measure or estimate the interference temperature, since cognitive users have difficulty in distinguishing between actual signals from a PU and noise or interference. Also, with the increase in the interference temperature limit, the SNR at the primary receiver decreases, resulting in a decrease in the primary network's capacity and coverage.

Spectrum sensing can be *non-cooperative* or *cooperative*. Cooperative sensing is theoretically more accurate because the uncertainty in a single user's detection can be minimized through collaboration. Moreover, multipath fading and shadowing effects can be mitigated so that the detection probability is improved in such an environment. However, the cooperative approaches cause adverse effects on resource-constrained networks due to the overhead traffic required for cooperation.

Spectrum sensing can be either *reactive* or *proactive*. With reactive sensing, a cognitive user performs spectrum sensing only when the cognitive user needs to access the spectrum (i.e., on-demand basis). With proactive sensing, a cognitive user continuously senses the spectrum. The selection between these spectrum sensing methods depends on the application requirements. For a delay-sensitive application, proactive spectrum sensing is preferred. However, for an energy-constrained application, reactive spectrum sensing could reduce energy consumption since sensing activities, which may consume a considerable amount of energy, can be minimized.

It should be noted that the identification of spectrum opportunities is not limited only to spectrum sensing performed by the cognitive users. Other possible approaches to identifying spectrum opportunities are database registry and beacon signals [6, Chapter 7]. For example, the application of a database registry, i.e., a radio environment map (REM), is described in [12, Chapter 11]. However, these approaches suffer from high infrastructure cost and may be infeasible for some application scenarios.

4.2.2.2 Spectrum analysis

In spectrum analysis, information from spectrum sensing is analyzed to gain knowledge about the spectrum holes (e.g., interference estimation, duration of availability, probability of collision with a licensed user due to sensing error, etc.), i.e., spectrum analysis is performed to estimate spectrum quality [6, Chapter 2]. One of the issues here is how to quantify the quality of the spectrum opportunity. The quality can be characterized by the SNR of the target frequency band, the average duration of the spectrum holes, the correlation of the availability of spectrum holes, etc. The information about the spectrum quality available to a cognitive user can be imprecise and noisy. The application of learning algorithms from artificial intelligence is one of the candidate techniques that could be employed by the CR network for spectrum analysis.

4.2.2.3 Spectrum decision

Based on the spectrum analysis, a spectrum decision to access the spectrum (e.g., frequency, bandwidth, modulation mode, transmit power, and time duration) is made by optimizing the system performance given the desired objective (e.g., maximize the throughput of the cognitive users) and constraints (e.g., maintain the interference caused to licensed users below the target threshold) [6, Chapter 2]. The spectrum decision does not depend only on the spectrum availability, but is also determined based on internal (and possibly external) policies [8]. The complexity of the decision model depends on the parameters considered during spectrum analysis (e.g., the average duration of the spectrum holes, the SNR of the target frequency band, the utility of the cognitive user obtained through accessing the spectrum holes). The decision model becomes more complex when a cognitive user has multiple objectives. Stochastic optimization methods (e.g., Markov decision process) would be an attractive tool to model and solve the spectrum access decision problem in a CR environment.

4.2.2.4 Spectrum sharing

After a decision on spectrum access has been made based on the spectrum analysis, the spectrum holes are accessed by cognitive users [6, Chapter 2]. Spectrum access is coordinated and managed by a cognitive medium access control (CMAC) protocol, which intends to avoid collisions with licensed users and also with other cognitive users. A CMAC protocol could be based on a fixed allocation MAC (e.g. FDMA, TDMA, CDMA) or on a random access MAC (e.g. ALOHA, CSMA/CA) [4]. Spectrum sharing should include much of the general functionality of any MAC protocol [8]. However, the unique characteristics of CR networks, such as the coexistence of cognitive users with licensed users and the wide range of available spectrum, impose additional specific requirements. Spectrum sharing can be centralized or distributed, cooperative or non-cooperative. For further information on spectrum sharing in regard to CMAC protocols, the interested reader is referred to [11, Chapter 7].

4.2.2.5 Spectrum mobility

Spectrum mobility is a function related to the change of the operating frequency band of cognitive users. When a licensed user starts accessing a radio channel which is currently being occupied by a cognitive user, the cognitive user must immediately vacate this channel and possibly move to an idle spectrum band. This change in the operating frequency band is referred to as spectrum mobility (aka spectrum handoff or spectrum handover). The objective of spectrum handover is to try to ensure that the transmission of the cognitive user can continue in the new spectrum band with minimum performance degradation. Since the latency due to spectrum handover could be high, cross-layer modification and/or adaption of components in the protocol stack may be required in order to achieve efficient QoS provisioning in the CR network. When a cognitive user performs spectrum handover, the receiver of the corresponding cognitive link must be notified of the spectrum handover (i.e., the synchronization requirement). Therefore, the CMAC protocol must be designed with provision for spectrum handover information exchange.

4.2.3 QoS Provisioning in CR Networks

The capacity evaluation and the QoS provisioning in CR networks are challenging and demanding tasks due to the changeable character of the spectrum available to the cognitive SUs. Since SUs dynamically use unoccupied spectrum on an opportunistic basis, the capacity of the CR network is inconstant and depends on the momentary spectrum utilization of the primary network. In order to provide significant capacity for the SUs, CR networks should be deployed in primary networks whose spectrum resources are sufficiently underutilized.

The QoS provisioning of the SUs is a complex task because a lot of requirements have to be taken into account. The QoS-related parameters of the secondary CR networks include (but are not limited to) the SU call blocking probability, the SU call dropping probability, the maximum tolerable transmission delay in the CR network, the average (or minimum guaranteed) throughput of a cognitive user, etc. Moreover, predefined interference requirements imposed by the primary network must be observed. Therefore, in order to achieve efficient QoS provisioning in CR networks, many design characteristics and cross-layer interdependencies should be taken into account.

There are numerous publications in the literature on various issues relevant to the QoS provisioning of CR networks for DSA (e.g., see [13–18] and the references therein). The throughput and the capacity of CR networks are investigated in [19–21] and in [22–24], respectively. The delivery of multimedia services over CR is studied in [25–27]. The call-level traffic capacity for VoIP SUs in a cognitive system with VoIP PUs is studied in [28]. An elaborate cross-layer analytical model for extensive analysis of the SU QoS provisioning and for thorough performance evaluation of CR networks is proposed in [29]. Channel reservation as a means of call-level SU QoS provisioning is investigated in [30].

In this section, channel limitation is analyzed as a means of call-level QoS provisioning in secondary CR networks operating in accordance with the hierarchical spectrum overlay approach for DSA [31]. Furthermore, the call-level performance of the CR network under consideration is investigated. The serving system is modeled by a two-dimensional state-transition diagram and a novel approximate but computationally efficient analytical approach for solving the state probabilities of the state-transition diagram is developed. A new precise formula for evaluation of the call dropping probability of the SUs is derived. Extensive simulation experiments are performed to validate the proposed analytical solutions. Numerical results are presented and some insightful conclusions are drawn.

4.2.3.1 The system model

In our study, we assume that the primary network and the cognitive network provide multimedia services with different bandwidth demands and that the bandwidth of a PU call is k times greater than the bandwidth of a SU call. Let us define the term *channel* as the necessary mean bandwidth for a PU multimedia call to be served and the term *subchannel* as the necessary mean bandwidth for a SU multimedia call to be served. It is obvious that one channel comprises k subchannels. The total bandwidth of the serving system is assumed to comprise n channels and hence nk subchannels.

We denote with i (i = 1, ... , n) and j (j = 1, ... , nk) the number of PU and SU calls in the system, respectively. The PU and SU call arrival streams are modeled by two Poisson random processes with arrival rates λ_p and λ_s, respectively. The PU and SU call durations follow negative exponential distributions with mean $1/\mu_p$ and $1/\mu_s$, respectively. Consequently, the offered PU traffic is $A_p = \lambda_p/\mu_p$, and the offered SU traffic is $A_s = \lambda_s/\mu_s$.

PUs have a preemptive priority over SUs. If a PU starts transmitting on a channel, all subchannels occupied by SUs within that channel have to be vacated immediately. If a channel is being used by a PU, the subchannels within that channel are unavailable to SUs. The service of PU calls is independent of the service of SU calls.

In the system model, perfect spectrum sensing, spectrum sharing, and spectrum handover procedures are assumed. Under these conditions, SU call blocking occurs only if there is not an unoccupied subchannel in the system to serve a new SU call. Similarly, SU call dropping occurs only if there is not an unoccupied subchannel in the system to continue the service of a SU call during spectrum handover. Since one channel comprises k subchannels, up to k SU calls can be dropped simultaneously at the arrival of a new PU call.

We model the described serving system via a two-dimensional continuous time Markov chain, as shown in Figure 4.1. Let us denote with $P_{i,\,j}$ the probability that the system is in state (i, j), i.e., the steady state probability that there are i PU calls and j SU calls in the system. Based on the state-transition diagram in Figure 4.1, we can derive the global balance equations:

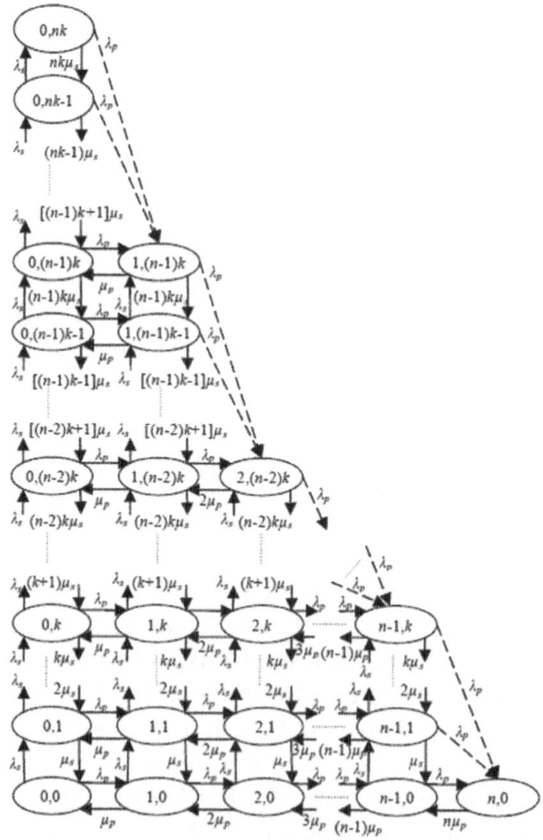

Figure 4.1 The state-transition diagram of the serving system.

$$(\lambda_p + \lambda_s)P_{0,0} = \mu_p P_{1,0} + \mu_s P_{0,1}; \tag{4.1}$$

$$(\lambda_p+\lambda_s+i\mu_p)P_{i,0} = \lambda_p P_{i-1,0}+(i+1)\mu_p P_{i+1,0}+\mu_s P_{i,1}, \text{where } 0 < i < n; \tag{4.2}$$

$$n\mu_p P_{n,0} = \lambda_p \sum_{j=0}^{k} P_{n-1,j}; \tag{4.3}$$

$$(\lambda_p + \lambda_s + j\mu_s)P_{0,j} = \lambda_s P_{0,j-1} + \mu_p P_{1,j} + (j+1)\mu_s P_{0,j+1}, \quad \text{where } 0 < j \leq (n-1)k; \tag{4.4}$$

$$(\lambda_p + \lambda_s + j\mu_s)P_{0,j} = \lambda_s P_{0,j-1} + (j+1)\mu_s P_{0,j+1}, \quad \text{where } (n-1)k < j < nk; \tag{4.5}$$

$$(\lambda_p + nk\mu_s)P_{0,nk} = \lambda_s P_{0,nk-1}; \tag{4.6}$$

$$(\lambda_p + i\mu_p + j\mu_s)P_{i,j} = \lambda_s P_{i,j-1} + \lambda_p \sum_{m=j}^{j+k} P_{i-1,m}, \quad \text{where } i > 0; j > 0; ik + j = nk; \tag{4.7}$$

$$(\lambda_p + \lambda_s + i\mu_p + j\mu_s)P_{i,j} = \lambda_p P_{i-1,j} + \lambda_s P_{i,j-1} + (j+1)\mu_s P_{i,j+1}, \quad \text{where } i > 0; j > 0; (n-1)k < ik + j < nk; \tag{4.8}$$

$$(\lambda_p+\lambda_s+i\mu_p+j\mu_s)P_{i,j} = \lambda_p P_{i-1,j}+\lambda_s P_{i,j-1}+(i+1)\mu_p P_{i+1,j}+(j+1)\mu_s P_{i,j+1}, \quad \text{where } i > 0; j > 0; ik + j \leq (n-1)k; \tag{4.9}$$

$$\sum_{i=0}^{n} \sum_{j=0}^{(n-i)k} P_{i,j} = 1. \tag{4.10}$$

The system of (4.1) – (4.10) contains $(n + 1)(nk + 2)/2$ unknown state probabilities and can be solved using an appropriate iterative method, such as the Gauss-Seidel method or the method of successive over-relaxation (SOR) (see [32, Chapter 4] for further information about these methods). However, due to the high computational complexity of these iterative methods, their implementation in CR may be infeasible with respect to the real-time and power consumption design requirements on CR.

4.2.3.2 The novel analytical approach

In this subsection, a new approximate approach for solving the state probabilities of the state-transition diagram (Figure 4.1) with relatively low computational complexity is presented. The accuracy of the proposed approach has been validated by extensive simulation experiments. Moreover, the approach is applicable in case that channel limitation is applied in the CR network. A new precise formula for evaluation of the SU call dropping probability is presented as well.

Channel limitation sets an upper bound on the maximum admissible number of SU calls in the CR network, i.e., it sets the SU call admission control (CAC) threshold. Let us denote with l the maximum allowable number of SU calls. If channel limitation is applied, $0 \leq j \leq l < nk$; otherwise: $0 \leq j \leq l = nk$.

Because of the preemptive priority of the PUs over the SUs, which is modeled by unidirectional transitions, the state-transition diagram (Figure 4.1) is not reversible. It differs from an ordinary reversible multidimensional state-transition diagram and a trivial solution based on state-based algorithms or the convolutional algorithm (see [33, Chapter 7]) cannot be applied.

Since the service of PUs is independent of the service of SUs and the PU and SU call arrival processes are i.i.d, we have:

$$P_{i,j} = P_i P_j^i = \frac{\frac{A_p^i}{i!}}{\sum_{m=0}^{n} \frac{A_p^m}{m!}} P_j^i, (i < n) \tag{4.11}$$

and

$$P_{n,0} = \frac{\frac{A_p^n}{n!}}{\sum_{m=0}^{n} \frac{A_p^m}{m!}} = B, \tag{4.12}$$

where P_j^i is the conditional probability that there are j SU calls in the system, provided that the number of PU calls is i; B is the PU call blocking probability.

Now we proceed to derive P_j^i by inspecting the columns of the diagram in Figure 4.1 and considering limitation if applied. Let us introduce the notations:

$$t = (n - i - 1)k, (0 \leq i < n) \tag{4.13}$$

and

$$\rho = \frac{\mu_p}{\mu_s}. \tag{4.14}$$

In states (i, j), where $i < n$ and $j > t$, SU call dropping occurs if a new PU call arrives, which is designated with the dashed transitions λ_p in Figure 4.1. Since dropping decreases the number of SU calls in the system, the effective departure (service) rate of SU calls in state (i, j) $\{i<n; j>t\}$, i.e., the rate of the transition from state (i, j) into state $(i, j-1)$, is assumed to be $j\mu_s + \frac{\lambda_p}{j-t}$ and local balance is assumed to exist between these two states. Based on these assumptions and taking into account the optional use of limitation $(0 < l \leq nk)$, we solve the balance equations about column i of the state-transition diagram for SU traffic only. Thus, we obtain:

$$P_j^i = \frac{\frac{A_s^j}{j!}}{\sum\limits_{m=0}^{l} \frac{A_s^m}{m!}}, t \geq l; j \leq l; i < n; \tag{4.15}$$

or

$$P_j^i = \frac{\frac{A_s^j}{j!}}{\sum\limits_{m=0}^{t} \frac{A_s^m}{m!} + \frac{1}{t!} \sum\limits_{x=t+1}^{\min[l,(n-i)k]} \frac{A_s^x}{\prod\limits_{m=t+1}^{x} \left(m+\rho \frac{A_p}{m-t}\right)}}, t < l; j \leq t; i < n; \tag{4.16}$$

or

$$P_j^i = \frac{\frac{A_s^j}{t! \prod\limits_{m=t+1}^{j} \left(m+\rho \frac{A_p}{m-t}\right)}}{\sum\limits_{m=0}^{t} \frac{A_s^m}{m!} + \frac{1}{t!} \sum\limits_{x=t+1}^{\min[l,(n-i)k]} \frac{A_s^x}{\prod\limits_{m=t+1}^{x} \left(m+\rho \frac{A_p}{m-t}\right)}}, t < l; t < j \leq \min [l, (n-i)k]; i < n. \tag{4.17}$$

Substituting (4.15) or (4.16) or (4.17) into (4.11), $P_{i,j}$ is obtained.

The main advantage of the analytical approach presented herein over the precise iterative methods for solving the state probabilities of the state-transition diagram of the system model is its computational simplicity which facilitates its application and implementation in CR systems with respect to satisfying the real-time and power consumption design requirements on CR.

SU call blocking occurs if all subchannels are occupied:

$$P_b = \sum_{i=0}^{n} P_{i,\min[l,(n-i)k]}. \tag{4.18}$$

SU call dropping occurs only if $j > t$ and a new PU call arrives. We present a new precise formula which evaluates the SU call dropping probability as the ratio of the mean number of dropped SU calls to the mean number of SU calls in the system (instead of just summing up the probabilities of the states where SU call dropping can occur):

$$P_d = \frac{\sum_{i=0}^{n-1} \sum_{j=t+1}^{\min[l,(n-i)k]} (j-t)\left[1-\exp\left(-\frac{\rho A_p}{j}\right)\right] P_{i,j}\Delta(l,t)}{\sum_{i=0}^{n-1} \sum_{j=1}^{\min[l,(n-i)k]} jP_{i,j}}, \tag{4.19}$$

where $\Delta(l,t) = 1$ if $l > t$, and $\Delta(l,t) = 0$ otherwise.

4.2.3.3 Analysis of channel limitation

In this subsection, the call-level performance of the CR network under consideration is investigated and the effect of applying channel limitation on the QoS provisioning of the SUs is analyzed. The call-level QoS provisioning in the CR network is investigated in terms of the SU call blocking probability P_b and the SU call dropping probability P_d. Numerical results are presented and insightful conclusions are drawn.

We first analyze the effect of the offered PU traffic load A_p on the performance of the secondary CR network. As A_p increases, both the SU call blocking probability P_b and the SU call dropping probability P_d increase, as shown in Figure 4.2 and Figure 4.3. This means that when the offered PU traffic increases, the traffic capacity of the CR network has to be reduced in order to maintain certain predefined QoS requirements with regard to the SU call blocking probability and the SU call dropping probability. Therefore, it is reasonable to deploy secondary CR networks for DSA only in primary networks whose spectrum resources are sufficiently underutilized.

Next, we analyse the effect of applying channel limitation on the QoS provisioning in the CR network. Figure 4.2 and Figure 4.3 illustrate that by applying limitation it is possible to reduce the SU call dropping probability P_d at the price of increasing the SU call blocking probability P_b, i.e., a trade-off relationship is observed between these two QoS parameters. As a general rule, users are much more intolerable to interruption (dropping) of an ongoing call than to rejection (blocking) of a new call. Hence, the use of limitation is favorable since the reduction in the SU call dropping probability P_d improves and facilitates the QoS of the SU calls already admitted to the CR network. Channel limitation is an efficient means of maintaining the SU call dropping probability P_d much smaller than the SU call blocking probability P_b, i.e., channel limitation is an efficient means of call-level QoS provisioning in the CR network.

Figure 4.2 and Figure 4.3 also illustrate that the effect of limitation is noticeable only when the offered PU traffic A_p is relatively small. If the offered PU traffic A_p is relatively large, the performance of the CR network is no longer determined by the SU CAC threshold l but by A_p, since, in this case, the number of PU calls in the system (and not l) limits the maximum possible number of SU calls that can be admitted and served. Consequently, in order to apply limitation effectively and to guarantee the QoS provisioning in the CR network, the PU traffic load must be relatively small, i.e., we again come to the conclusion that the primary network has to be relatively underutilized in order to provide QoS-guaranteed services over CR.

Figure 4.4 shows that the traffic capacity A_s of the secondary CR network decreases as the SU CAC threshold l decreases, i.e., limitation reduces the capacity of the CR network. Therefore, limitation should be prudently applied

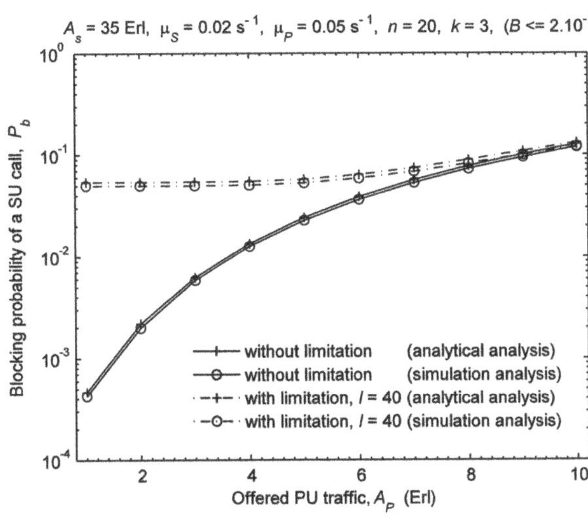

Figure 4.2 SU call blocking probability versus the offered PU traffic.

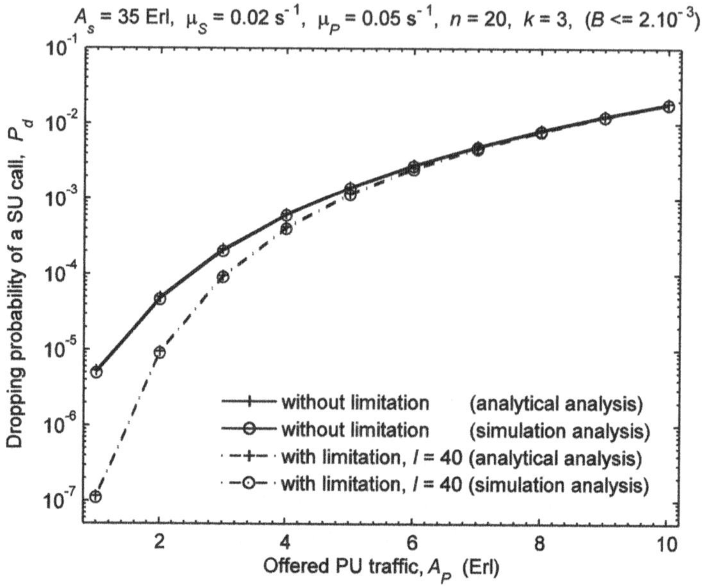

Figure 4.3 SU call dropping probability versus the offered PU traffic.

in order to satisfy the QoS requirements of the SUs without excessively decreasing the traffic capacity of the CR network.

As shown in Figure 4.2, Figure 4.3, and Figure 4.4, various simulation experiments have been performed and in all cases a good coincidence between analytical and simulation results is observed. The analytical analysis is completely consistent with the simulation analysis, which validates and verifies the developed novel approximate approach for solving the state probabilities of the state-transition diagram of the system model under consideration.

Our study corroborates that the effect of the PU traffic on the capacity and performance of the CR network must always be considered. CR should be used in underutilized primary networks. Since the transmission resources available to the CR network are determined by the momentary PU activity and traffic load, the QoS provisioning in the secondary CR network is a challenging task. Channel limitation decreases the capacity of the CR network but facilitates its QoS provisioning by reducing the SU call dropping probability. The presented analytical model may provide guidelines for the design and performance optimization of secondary CR networks for DSA.

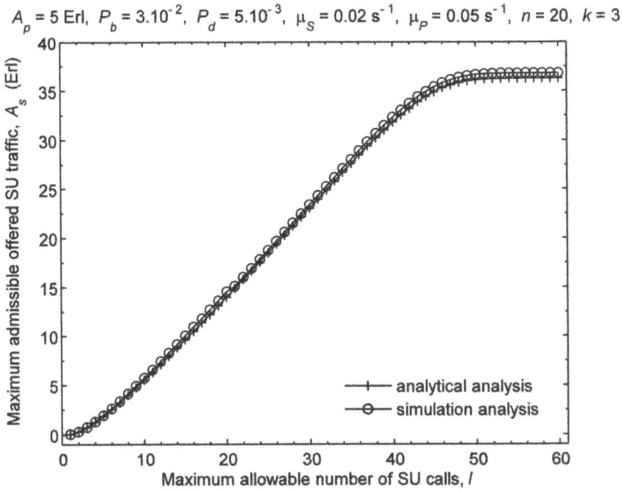

Figure 4.4 Cognitive traffic capacity versus the SU CAC threshold.

4.3 Performance Analysis of AMC-enabled Wireless Access Networks

Today's Internet concept and communications infrastructure need to be adapted to the growing demands of various kinds of services and business models. Development of the network of the future is a significant step towards achievement of this objective, taking advantages of the latest technologies. Within this scope, wireless communications play a key role due to the benefits made available to end users by evolution of wireless access technologies over several generations. This is accompanied by adaption of advanced technologies, at different layers of system model, in order to achieve enhanced system performance. For proper dimensioning of network resources new models should be developed, which match to the wireless systems principles of operation.

In order to overcome the limitation of wireless environment, significant research efforts have been put towards technologies, which enhance spectral efficiency. Adaptive Modulation and Coding (AMC) has been well adopted as an advanced physical layer technology for that purpose. From teletraffic point of view, in AMC-enabled systems the bandwidth demands of active connections may fluctuate due to movement of mobile stations (MS) within the cell coverage. This is true for services requiring constant transmission capacity. The resulting intra-cell handover leads to a reallocation of resource amount placed to the connection disposal and has a major impact on the system performance. In the presence of mobile users willing to access to a rich variety of services, advanced mobility management is needed such that these services are provided seamlessly. Wireless resource management policies should be capable of determine the optimal use of limited and expensive resources, according to the channel state information and target QoS requirements.

In the context of resource management, the problem is directly related to the issue of resource reservation and admission control. This means each AMC-enabled base station (BS) should be equipped with a threshold-based bandwidth management policy for QoS guarantee, which takes into consideration intra-cell handover influence on resource consumption.

4.3.1 Handover Management in Wireless Access Networks

For efficient usage of limited resources, wireless communications systems employ the cellular concept. In such a multicell environment new problems arise. The following two aspects could be distinguished – inter-cell interference management and handover management. Since network resources are dynamically loaded, as a consequence of users' mobility, the attention is paid to a handover management policy, which maintains target QoS of active connections, over new ones. It aims at balancing the QoS of both handover and new calls (connections) arrivals. This is done by employing an appropriate resource reservation and admission control mechanisms.

Wireless resource management policies should be capable of determine the optimal use of limited resources, according to the wireless channel state information and specific QoS requirements. Since bandwidth is a fundamental wireless resource, from teletraffic point of view, it may be considered as a transmission capacity of the wireless medium, which is shared by multiple users.

Initially, handover management for single-service wireless access systems had been extensively studied. In legacy wireless access systems each active connection is assigned a fixed amount of bandwidth independently of the MS location within a cell, as the quantity of resource consumption depends on the traffic source characteristics. A number of prioritization policies that handle the handover traffic has been proposed and analyzed (see for example [34], [35] and the references therein). The basic mechanism is to suitably partition the available resource in a cell to a different traffic types (new and handover call arrivals). The investigations are made under the assumption of conversational services, such a voice, occupying a fixed number of resource units (channels) per call. The essence of admission control policy is to set a higher priority on handover arrivals (an active MS crossing cell boundary) over newly generated calls within a cell in order to satisfy stringent QoS requirements of handover traffic. Dropping a handover connection has a severe impact on users' satisfaction than blocking a newly originated call. Such a prioritization policy can be implemented by a fixed number of channels (resource units), referred to as "guard channels", exclusively allocated for handover call arrivals. The basic performance metrics, such as handover calls dropping probability and blocking probability of newly originated calls within a cell may be adjusted by setting an appropriate admission threshold.

A common feature of admission policies is their ability to use the target cell information only (i.e., the number of occupied channels). It would be more beneficial some additional information of the adjacent cell to be used. Since handover decision is solely based on the availability of a free channel without taking into consideration the signal quality, [36] extends the well-known "guard channel" (GC) policy. This is done by combining the mobile assisted handover technique (available at GSM cellular system) and GC. A more efficient handover management scheme is proposed in [37]. The future behavior of an active call can be estimated more precisely, based on the mobility (location, direction and velocity) of each MS. A channel reservation/cancellation message can be sent to the approaching cell, allowing adaptive resource reservation where each BS dynamically adjusts the admission threshold according to these messages.

In the meantime, the challenges in handover management schemes are towards efficient sharing of scarce resources among multiple traffic classes in order for the emerging wireless access systems to support multimedia services. These are characterized with particular QoS and bandwidth requirements [38]. An attention is paid to handover management of each traffic class to deal with the different QoS requirements – each traffic class encompass two kinds of arrivals (new and handover) and this results in different admission threshold to be set. The overall cell resource, in terms of resource units, is divided into several regions. In general, the following resource management approaches can be distinguished – complete partition (reservation of a bandwidth exclusively for each traffic class); complete sharing (sharing of available bandwidth among all traffic classes). For a particular traffic class, it is possible a limitation level to be set for the purpose of managing traffic congestion. There could also exist hybrid schemes, which combine the above-mentioned approaches (this issue was initially arisen in service-integrated networks, e.g., ISDN – [33]). A comparative study of the first two schemes is carried out in

[39]. It is shown the advantage of complete sharing to complete partition scheme for efficient use of scarce resource in case of two traffic classes with different bandwidth requirements. The potential of movable boundaries allocation strategies, that can adjust the number of channels for each traffic class, has been investigated in [40]. The proposed scheme extends the GC by introducing different admission thresholds for different traffic classes. It is assumed that each traffic class requires one channel (resource unit) per connection. A further extension of bandwidth allocation schemes should study the policy performance, supporting traffic flows with variable bandwidth requirements, which makes it possible more accurate estimates of resource quantity [41].

In wireless networks the channel capacity of a wireless link is time-varying, and thus QoS requirements may not be satisfied, even though a large amount of resource (i.e., bandwidth) is allocated to a certain connection. This is especially true when a MS is located near the cell border area. In order for the emerging wireless access systems to overcome the limitation of wireless communications environment and to maintain a target packet error rate (PER) over wireless links, significant research efforts have been put towards development of technologies for enhancing the spectral efficiency. AMC has been well adopted as an advanced physical layer technology to match transmission parameters to time-varying channel conditions [42]. In contrast to the legacy wireless access systems, where each active connection is assigned a fixed amount of bandwidth during the whole connection time (independently of the MS location within a cell), in AMC-enabled systems (cells) the bandwidth demands of active connections may vary due to movement of MS within a cell coverage – each MS is served by a particular modulation and coding scheme (MCS), depending on the radio channel conditions. This is true for services requiring constant transmission capacity. As a consequence, the resulting intra-cell handover leads to reallocation of resource amount necessary for an active connection and its influence on the system performance should be taken into consideration.

Considering the importance of the problem, a great extent of research work has been devoted to development of handover management policies applied for the case of inter-cell handover only. Having interested in performance analysis of emerging wireless access technologies, the application of above-mentioned methods may be inaccurate. This requires AMC schemes to be integrated into analytical models for solving problems of this kind [43–45]. So far the problem of intra-cell handover influence has been covered in a number of research activities, which mainly focus on the physical layer performance, i.e., considerable amounts of interference is suppressed by means of intra-cell handover. For instance, an interference avoidance technique based on the use of intra-cell handover in OFDMA femtocells is proposed in [46] where the concept of intra-cell handover in GSM networks is applied to OFDMA subchannels selection, in order to mitigate the interference between macro- and femtocell tiers. For proper resource dimensioning and management in AMC-enabled wireless systems it is necessary intra-cell handover influence on the data link layer performance to be taken into account.

4.3.2 Intra-cell Handover Management in AMC-enabled Wireless Access Networks

Let us consider a cell with a fixed amount of resource. Independently of the multiple access technology employed, the cell capacity could be represented in terms of effective bandwidth. Thus, the total cell capacity is C resource units (bandwidth units).

An AMC scheme is employed at the physical layer to match the transmission rate to the time-varying channel conditions as well as to maintain a target PER over wireless link. This results in splitting up the overall cell area on several regions (rings), corresponding to the available modes of AMC scheme, each of which represents a pair of specific modulation format and a forward error correcting (FEC) code (Figure 4.5).

There exist different classes of traffic, each of which requires different amount of resource units per connection, based on the AMC employment. A call, with certain traffic characteristics, to (from) a MS in Ring l (l = 1, L)

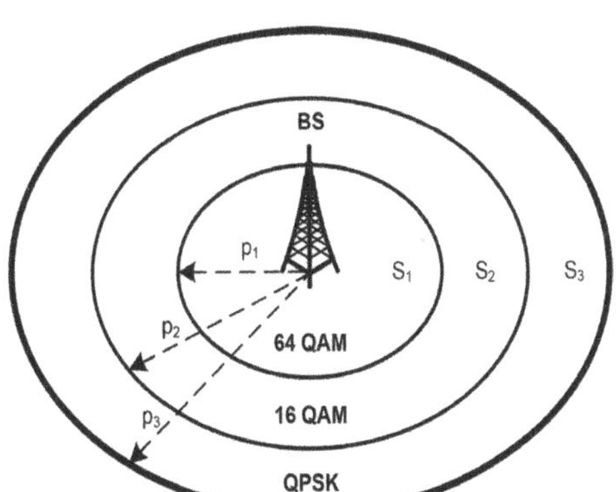

Figure 4.5 An AMC-enabled cell of wireless access network.

simultaneously requires d_l resource units in order to be guaranteed desired QoS. The rings are numbered starting with the most inner being number 1 to the most outer cell ring being number L. Therefore, $d_x > d_y$, if $x > y$.

A new call arrival in both rings follows a Poisson process with mean rate λ_l ($l = 1, L$). For the sake of simplicity it is assumed that both new call duration and Ring l dwell time are negative exponentially distributed random variables with mean $1/\mu$ and $1/\delta_l$, respectively.

When a MS moves within a cell coverage it crosses inner boundaries, defined by particular MCS, resulting in intra-cell handover occurrence. The intra-cell handover rates δ_l are related to the MSs average speed v, rings area S_l, and the MSs distribution within the cell [47], [48]. δ_l^+ denotes a call handover rate outward of Ring l (from Ring l to Ring l+1), while δ_l^- – from Ring l to Ring l–1. Bandwidth demand of an active connection is not constant any more, which could lead to QoS degradation in case of insufficiency of transmission resources. To maintain the QoS of each user to an acceptable level, a threshold-based bandwidth management policy in AMC-enabled system is required. It shall prioritize connections (both new and handover call arrivals) according to the target QoS requirements and the cell area of a call origin.

The threshold-based bandwidth management policy is based on the complete sharing approach (Figure 4.6) and it is applicable for a cell employing 64-QAM and 16-QAM modulation schemes at physical layer. Both new calls (NC1) and handover calls (HO1) incoming in the cell area served by the highest MCS get the highest priority (full access to the available cell resources). These calls are more profitable for the network operator, because they are allocated the smallest amount of resources per active connection. This area is referred to as “Ring 1”, linked with the 64-QAM modulation scheme. Both new calls (NC2) and handover calls (HO2) offered to “Ring 2” area compete for the remaining resource. This is done by setting the threshold levels T_0 and T_1, in terms of maximum number of resource units. When the number of occupied resource units is less than T_0, all traffic classes (both new and handover calls offered from both rings) can be admitted to the system (cell). Since dropping a handover connection (HO2) is not considered acceptable, $T_1 > T_0$. The policy aims at adjusting the threshold levels, in order to satisfy the handover dropping probability, keeping reasonable values of new calls blocking probabilities.

Under the assumptions stated above, the handover management method can be modeled as a multi-dimensional Markov process. The multidimensional random variable depends on the number of rings L in a cell. The system (cell)

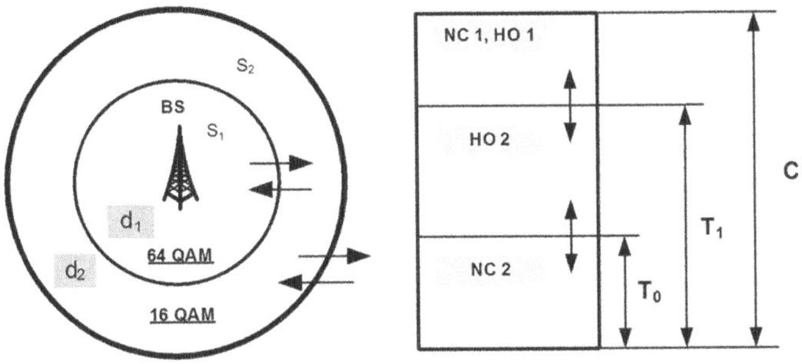

Figure 4.6 An AMC-enabled cell with a handover management policy.

state-space, from a data link perspective, is defined by the set

$$s = \{n_1, n_2, ..., n_l, ..., n_L\}, \tag{4.20}$$

where n_l denotes the number of active MSs in ring l, ($l = 1, L$).

Since we are interested in the steady state probabilities estimation, the challenge is to solve the system of linear equation for such a process. The possible transitions out of any state $\{n_l\}$ and into the same state $\{n_l\}$, due to changes in ring l, are shown on Figure 4.7. In a condition of statistical equilibrium the steady-state balance equations can be obtained. The topic under consideration faces the problem of irreversibility (i.e., the underlying Markov process is non-reversible). Thus, the steady states probabilities cannot be calculated by using product form solutions. A common approach is towards a direct method application, which requires a matrix inversion technique to be used. Depending on the structure of the coefficient matrix, an appropriate numerical method (algorithm) for solving simultaneous linear equations is required. This could limit the application of the method to relatively small state-spaces. The same could be true for a specific class of recursive methods for solving non-Markovian processes [49]. Although the method is well-adopted [41], [50], the limitation could arise in case a system with large state space and number of state transitions is investigated. In order to fully exploit the advantages of certain recursive methods applied to reversible Markov processes, a challenging work is to construct a reversible Markov chain that well approximate the non-reversible system under consideration (in a manner similar to [51]).

If it is assumed $L = 2$ (Figure 4.6), the call (connection) level performance measures of interest are obtained by solving a two-dimensional non-reversible Markov process.

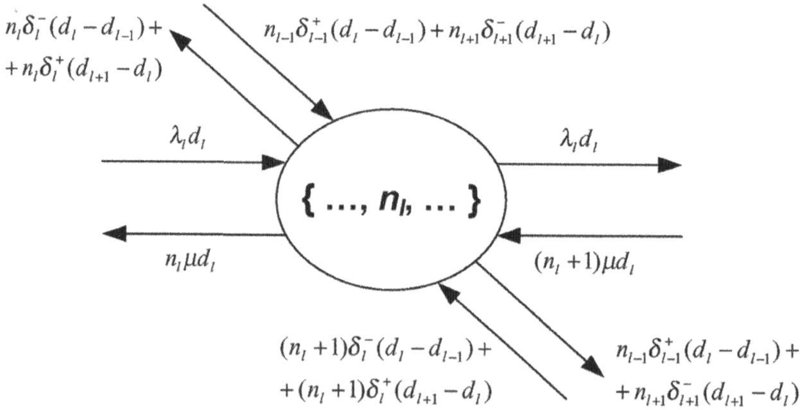

Figure 4.7 System state transitions for Ring *l*.

The set of allowable states, denoted by *S*, is determined by the resource sharing policy in use. Therefore, $s \in S$ if and only if the following conditions are satisfied

$$S = \{s : 0 \leq n_2 \cdot d_2 \leq T_0 \leq T_1;\ 0 \leq \sum_{i=1}^{2} n_i \cdot d_i \leq C\}. \tag{4.21}$$

Further, the following notation is introduced $s_l^+ = \{n_1, n_2, ..., n_l + 1, ..., n_L\}$.

The system performance measures can be obtained after calculating the steady-state probabilities. Both new (NC1) and handover (HO1) calls, offered to Ring 1, share the entire available cell resource. Since handover calls HO1 are offered from the Ring 2 area, served by a lower order MCS, these ones request smaller amount of resources to be served and QoS to be guaranteed. For this reason, HO1 arrivals (from Ring 2) cannot be dropped. The call blocking probability of new calls (NC1) in Ring 1 area can be derived as

$$P_{NC1} = \sum_{s \in BD_{NC1}^+} P(s), \tag{4.22a}$$

$$\begin{aligned} BD_{NC1}^+ &= \{s \in S | s_1^+ \notin S\} \\ &= \{s \in S | (d_1 + \sum_{j=1}^{2} n_j \cdot d_j) > C\}. \end{aligned} \tag{4.22b}$$

Let us introduce the subset $\hat{S} \in S$, defined by the threshold level T_0. The cell resource is completely shared by all types of calls offered from both rings. Thus, the blocking probability P_{NC2} for new calls offered to Ring 2 is given by

$$P_{NC2} = \sum_{s \in B^{+}_{NC2}} P(s) \tag{4.23a}$$

$$\begin{aligned} B^{+}_{NC2} &= \{s \in S | s^{+}_{2} \notin \hat{S}\} \\ &= \{s \in S | d_2 + n_2 \cdot d_2 > T_0 \vee T_0 < n_2 \cdot d_2 \leq T_1 \vee \\ &\vee d_2 + \textstyle\sum_{j=1}^{2} n_j \cdot d_j > C\}. \end{aligned} \tag{4.23b}$$

The dropping probability P_{HO2} for handover calls offered to Ring 2 is given by

$$P_{HO2} = \sum_{s \in D^{+}_{HO2}} P(s), \tag{4.24a}$$

$$\begin{aligned} D^{+}_{HO2} &= \{s \in S | s^{+}_{2} \notin S\} \\ &= \{s \in S | d_2 + n_2 \cdot d_2 > T_1 \vee d_2 + \sum_{j=1}^{2} n_j \cdot d_j > C\}. \end{aligned} \tag{4.24b}$$

Consider the case where the total cell capacity is 20 resource units. Based on the resource management policy available resource is shared by both new call arrivals from each ring, and intra-cell handover arrivals, as a result of MSs movement across inner cell boundaries (rings). It is assumed a cell radius r = 2 km. The relation of the outer radius of Ring 1, linked to a MCS of the highest order (64-QAM), to the cell radius is denoted by p. This proportion cannot be set and depends on the wireless environment propagation conditions and statistical characteristics. The distance covered by a MCS is also governed by a set of target performance measures at physical level (e.g., packet error rate, SINR, etc.). For constant transmission capacity provisioning, the necessary amount of resource units for type l call is based on the relation between MCSs employed (ratio of spectral efficiency values of MCSs for particular coding rates). For case considered here $d_1 = 1$, $d_2 = 2$ resource units per call [52]. In a similar manner, the new call arrival rate λ_l at Ring l is tightly coupled with the ratio of the Ring l area S_l to the entire cell area. Switching among MCS modes is governed by the wireless channel propagation conditions.

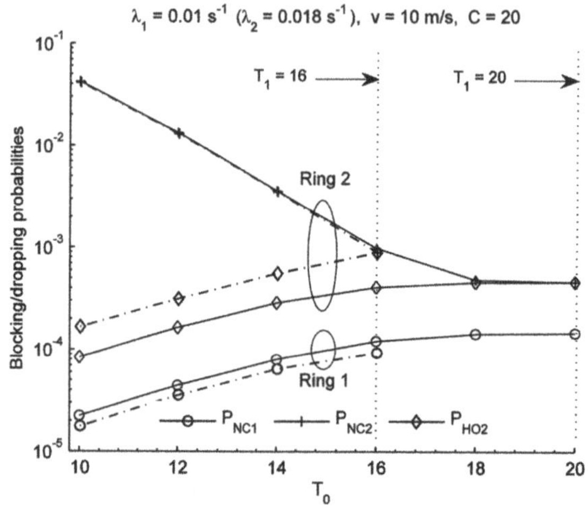

Figure 4.8 Influence of Ring 2 limitation thresholds on QoS metrics.

For complete sharing scenario ($T_0 = T_1 = 20$) both type of traffic flows (new and handover calls) offered from Ring 2 area experience higher losses compared to traffic arrivals from Ring 1 (Figure 4.8). Connections with lower bandwidth requirements per call, as those offered from MSs served by the highest order modulation scheme, have a better chance at occupying the available resource units than those with higher bandwidth requirements. In case handover limitation threshold T_1 is set, it is shown that significant improvement of the performance measures cannot be reached. It is more likely Ring 2 handover dropping probability to get increased. For this reason, it is more desirable handover requests to Ring 2 area to get full access to available resources.

Figure 4.9 depicts the impact of users' (MSs) mobility on the system performance. It can be seen that the resource management policy has low efficiency for high mobility users, compared to low mobility ones, as a result of increased intra-cellular handover rates. There exists a tradeoff between new call blocking probability and handover call dropping probability. For this reason, the threshold level T_0 has to be carefully set, such that Ring 2 new call blocking meet QoS.

The influence of the Ring 1 region area is illustrated on Figure 4.10 (in terms of the parameter p). The parameter cannot be set and it depends on the wireless environment conditions. The following observations can be made from the figure. Again, the resource management policy is efficient for

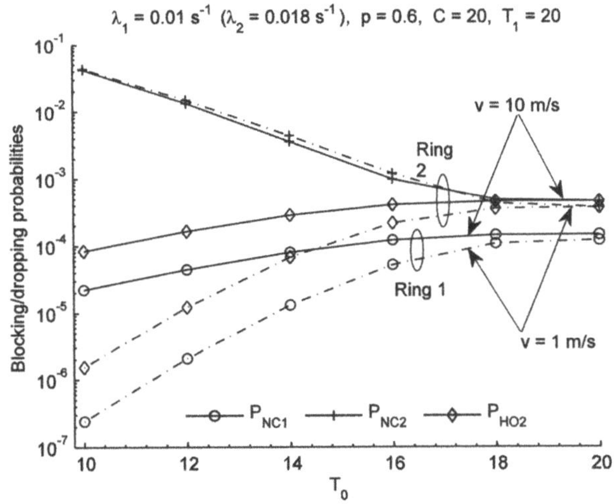

Figure 4.9 Influence of MSs mobility on the system performance.

low mobility users, but it is followed by certain limitations. If the wireless transmission experience good channel conditions, AMC uses the highest order modulation and it would lead to dominant influence on Ring 1 area in cell coverage (p is getting higher values). More significant part of the offered traffic flows is served most effectively (with the smallest amount of resources per active call) with minimal losses. This is a case when the "complete sharing" policy (full access to the available cell resources of all traffic classes) is applied. Implementation of "class limitation" policy, in such a case, is not recommended, because it would lead to minimal reduction of handover dropping probability, and significantly increase the blocking probabilities of new calls to unacceptable levels. A good resource management scheme has to balance the tradeoffs between new call blocking and handover dropping probabilities in order to meet target QoS requirements. An optimal point of operation of "class limitation" policy can be achieved when wireless environment provides such conditions that $p \approx 0,6$ (MSs falling in approx. 60 $\%$ of cell coverage are served by the highest order modulation – 64 QAM).

The policy operation under certain dynamic range of new calls arrival rates is depicted on Figure 4.11. The relation between the rings' new incoming call rates is based on the relations between the distances covered by MCSs employed. Again the policy takes effect on low mobility users. Based on the traffic load either "complete sharing (CS)" or "class limitation (Limitation)"

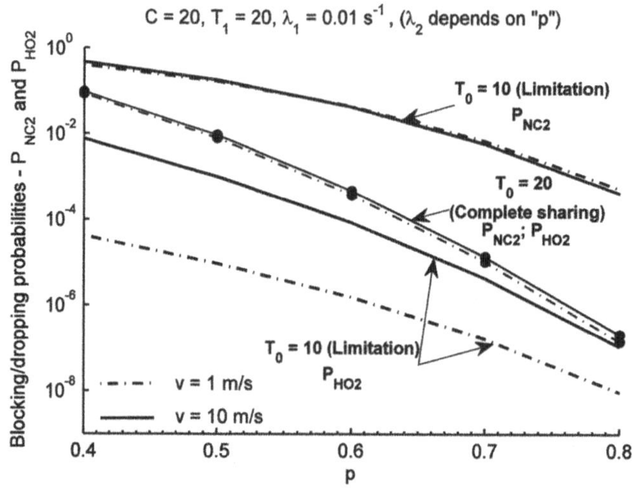

Figure 4.10 Ring 1 area influence on the system performance.

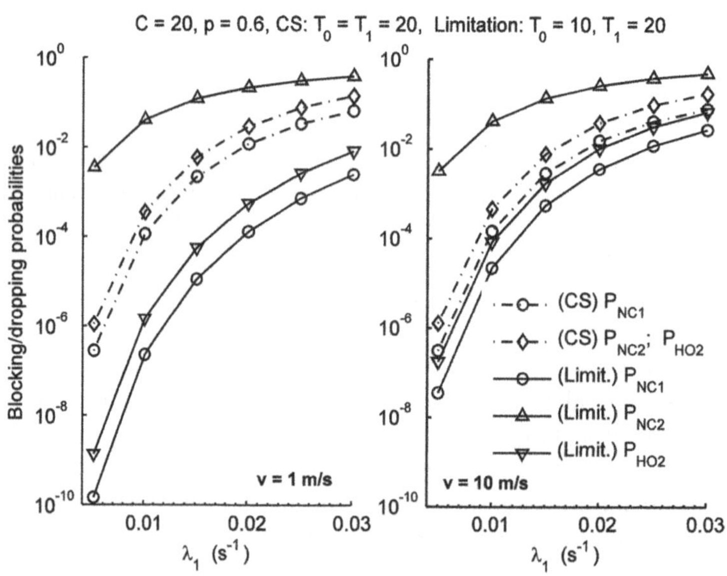

Figure 4.11 Call-level QoS parameters vs. new call arrival rate dynamic.

approach could be applied. For a given amount of resources into the cell, when the new call arrival rate is high, no matter how system parameters are adjusted, the policy cannot guarantee desired QoS.

4.4 Conclusions

Wireless communications brought the freedom of users to get an access to a rich variety of services, independently of their location. This benefit is made possible by successive evolution of wireless communication technologies. Throughout this evolution there still exist difficulties in satisfying users' desire for QoS provisioning, similar to that provided by the wired access technologies, due to limitations in wireless communications channels. CR and AMC has been adopted as two different advanced technologies to partially cope with these limitations.

In section 4.2, the application of CR networks for DSA has been discussed. A brief overview of the well-known approaches for DSA has been presented in subsection 4.2.1. Spectrum management functions essential to the operation of the CR networks for DSA have been described and explained in subsection 4.2.2. The QoS provisioning in CR networks has been addressed in details in subsection 4.2.3. Channel limitation has been analyzed as a means of call-level QoS provisioning in secondary CR networks operating in accordance with the hierarchical spectrum overlay approach for DSA.

The CR concept is expected to push efficiency in spectrum access and resource allocation beyond the traditional limits. The QoS provisioning in CR networks for DSA is a complex task of high importance. There are still many open research issues and design challenges in the fields of CR networks and DSA that need to be solved and overcome. Besides the technical challenges, business, regulatory, and political challenges have to be addressed as well.

In section 4.3 the application of AMC has been discussed. In contrast to the legacy wireless access systems, where to each active connection is assigned a fixed amount of bandwidth during the whole connection time, in AMC-enabled systems the bandwidth demands of active connections may vary due to movement of MS within a cell. This is true for services requiring constant transmission capacity. As a consequence, the resulting intra-cell handover leads to reallocation of resource amount necessary for an active connection.

The problem of intra-cell handover influence has been covered in a number of research activities, which mainly focus on the physical layer performance. For proper resource dimensioning and management in AMC-enabled wireless systems the intra-cell handover influence on the upper layers performance is taken into account. This results in AMC schemes integration into analytical models for solving problems of this kind. A threshold-based bandwidth management policy in AMC-enabled system is presented, which prioritize connections according to the target QoS requirements and the cell area of a call origin.

The use of some kind of resource reservation in both cases (CR and AMC) for call dropping probability reduction leads to state-transition diagram irreversibility (i.e., the underlying Markov process is non-reversible). Thus, the steady states probabilities cannot be calculated by using product form solutions. An approximate (section 4.2) and a direct equation solving (section 4.3) approaches are applied. There still exist open research issues in solving non-reversible Markov processes in case of complex systems with large state spaces and numbers of state transitions are investigated. A challenging work may focus on methods for reversible Markov chains construction, which well approximate the non-reversible system investigated. This will enable us to use attractive recursive methods (e.g., Kaufman-Roberts) that can be applied to realistic systems with large state-spaces.

References

[1] Zhao, Q., and B. Sadler, "A Survey of Dynamic Spectrum Access: Signal Processing, Networking, and Regulatory Policy," IEEE Signal Processing Magazine, Vol. 24, No. 3, May 2007, pp. 79–89.

[2] Mitola III, J., and G. Q. Maguire, "Cognitive Radio: Making Software Radios more Personal," IEEE Personal Communications, Vol. 6, No. 4, August 1999, pp. 13–18.

[3] Mitola III, J., "An Integrated Agent Architecture for Software Defined Radio," Ph.D. dissertation, Royal Institute of Technology (KTH), Sweden, 8 May 2000.

[4] Buddhikot, M. M., "Understanding Dynamic Spectrum Access: Models, Taxonomy and Challenges," The 2nd IEEE International Symposium on New Frontiers in Dynamic Spectrum Access Networks (IEEE DySPAN), Dublin, Ireland, April 2007, pp. 649–663.

[5] IEEE Std. 1900.1–2008, IEEE Standard Definitions and Concepts for Dynamic Spectrum Access: Terminology Relating to Emerging Wireless Networks, System Functionality, and Spectrum Management, September 2008.

[6] Hossain, E., D. Niyato, and Z. Han, Dynamic Spectrum Access and Management in Cognitive Radio Networks, New York, USA, Cambridge University Press, 2009.

[7] Shin, K. G., H. Kim, A. W. Min, and A. Kumar, "Cognitive Radios for Dynamic Spectrum Access: from Concept to Reality," IEEE Wireless Communications, Vol. 17, No. 6, December 2010, pp. 64–74.

[8] Akyildiz, I. F., W.-Y. Lee, M. C. Vuran, and S. Mohanty, "A Survey on Spectrum Management in Cognitive Radio Networks," IEEE Commun. Mag., Vol. 46, No. 4, April 2008, pp. 40–48.

[9] Doyle, L., Essentials of Cognitive Radio, New York, USA, Cambridge University Press, 2009.

[10] Chen, K.-C., and R. Prasad, Cognitive Radio Networks, Great Britain, John Wiley & Sons, 2009.

[11] Xiao, Y., and F. Hu (editors), Cognitive Radio Networks, Boca Raton, Florida, USA, CRC Press, 2009.

[12] Fette, B. (editor), Cognitive Radio Technology, 2nd ed., USA, Academic Press (Elsevier), 2009.

[13] An, C., H. Ji, and P. Si, "Dynamic Spectrum Access with QoS Provisioning in Cognitive Radio Networks," IEEE Global Telecommunications Conference (GLOBECOM), Miami, Florida, USA, December 2010, pp. 1–5.

[14] Alshamrani, A., X. S. Shen, and L.-L. Xie, "QoS Provisioning for Heterogeneous Services in Cooperative Cognitive Radio Networks," IEEE Journal on Selected Areas in Communications, Vol. 29, No. 4, April 2011, pp. 819–830.

[15] Li, J., T. Zhu, P. Xu, and T. K. T. To, "QoS Provisioning Multi-Level Spectrum Allocation Algorithm," International Conference on Advanced Technologies for Communications (ICATC), Ho Chi Minh City, Vietnam, October 2010, pp. 62–67.

[16] Hasan, R., and M. Murshed, "Provisioning Delay Sensitive Services in Cognitive Radio Networks with Multiple Radio Interfaces," IEEE Wireless Communications and Networking Conference (WCNC), Cancun, Mexico, March 2011, pp. 162–167.

[17] Canberk, B., I. F. Akyildiz, and S. Oktug, "A QoS-aware Framework for Available Spectrum Characterization and Decision in Cognitive Radio Networks," IEEE 21st International Symposium on Personal Indoor and Mobile Radio Commun. (PIMRC), Istanbul, Sept. 2010, pp. 1533–1538.

[18] Xin, Q., and J. Xiang, "Joint QoS-aware Admission Control, Channel Assignment, and Power Allocation for Cognitive Radio Cellular Networks," IEEE 6th International Conference on Mobile Adhoc and Sensor Systems (MASS), Macau, 12–15 Oct. 2009, pp. 294–303.

[19] Guo, W., and X. Huang, "Maximizing Throughput for Overlaid Cognitive Radio Networks," IEEE Military Communications Conference (MILCOM), Boston, MA, USA, October 2009, pp. 1–7.

[20] Lei, G., C. Hu, W. Wang, T. Peng, and W. Wang, "Increase the End-to-End Throughput of a Cognitive Radio Chain by Considering the Primary Usage Pattern and Transmission Scheduling," IEEE Wireless Communications and Networking Conference (WCNC), Budapest, Hungary, April 2009, pp. 1–6.

[21] Simeone, O., Y. Bar-Ness, and U. Spagnolini, "Stable Throughput of Cognitive Radios with and without Relaying Capability," IEEE Transactions on Communications, Vol. 55, No. 12, December 2007, pp. 2351–2360.

[22] Rashid, R. A., N. M. Aripin, N. Fisal, and S. K. S. Yusof, "Inner Bound Capacity Analysis of Cooperative Relay in Cognitive Radio using Information Theoretic Approach," IEEE 9th Malaysia International Conference on Communications (MICC), Kuala Lumpur, Malaysia, December 2009, pp. 328–331.

[23] Cho, H., and J. Andrews, "Upper Bound on the Capacity of Cognitive Radio without Cooperation," IEEE Transactions on Wireless Communications, Vol. 8, No. 9, Sept. 2009, pp. 4380–4385.

[24] Rezki, Z., and M.-S. Alouini, "On the Capacity of Cognitive Radio under Limited Channel State Information," 7th International Symposium on Wireless Communication Systems (ISWCS), York, UK, 19–22 Sept. 2010, pp. 1066–1070.

[25] Chaoub, A., E. Ibn Elhaj, and J. El Abbadi, "Multimedia Traffic Transmission over TDMA shared Cognitive Radio Networks with Poissonian Primary Traffic," International Conference on Multimedia Computing and Systems (ICMCS), Ouarzazate, Morocco, April 2011, pp. 1–6.

[26] Chen, Y., Y. Wu, B. Wang, and K. J. R. Liu, "Spectrum Auction Games for Multimedia Streaming over Cognitive Radio Networks," IEEE Transactions on Communications, Vol. 58, No. 8, August 2010, pp. 2381–2390.

[27] Shiang, H.-P., and M. van der Schaar, "Queuing-based Dynamic Channel Selection for Heterogeneous Multimedia Applications over Cognitive Radio Networks," IEEE Transactions on Multimedia, Vol. 10, No. 5, August 2008, pp. 896–909.

[28] Mihov, Y. Y., and B. P. Tsankov, "Cognitive System with VoIP Secondary Users over VoIP Primary Users," The Third International Conference

on Advanced Cognitive Technologies and Applications (COGNITIVE 2011), Rome, Italy, 25–30 September 2011, pp. 30–35.

[29] Mihov, Y. Y., "Cross-Layer Analysis and Performance Evaluation of Cognitive Radio Networks," The 6th International Conference on Systems and Networks Communications, (ICSNC 2011), Barcelona, Spain, 23–29 October 2011, pp. 99–104.

[30] Mihov, Y. Y., and B. P. Tsankov, "QoS Provisioning via Channel Reservation in Cognitive Radio Networks," The Third International IEEE Conference on Microwaves, Communications, Antennas and Electronic Systems (IEEE COMCAS 2011), Tel Aviv, Israel, 7–9 Nov. 2011, pp. 1–6.

[31] Mihov, Y. Y., and B. P. Tsankov, "Call-Level Performance Evaluation and QoS Provisioning in Cognitive Radio Networks," IEEE AFRICON 2011, Livingstone, Zambia, 13–15 September 2011, pp. 1–5.

[32] Cooper, R. B., Introduction to Queueing Theory, 2nd ed., USA, Elsevier North Holland, 1981.

[33] Iversen, V. B., Teletraffic Engineering and Network Planning, COM Department, Technical University of Denmark, May 2010.

[34] Ramjee R. et al., "On Optimal Call Admission Control in Cellular Networks", Journal on Wireless Networks, Kluwer Academic Publishers, Vol. 3, No. 1, 1997, pp. 29–41.

[35] Markoulidakis J. G. et al., "Optimal System Capacity in Handoff Prioritized Schemes in Cellular Mobile Telecommunication Systems", Journal on Computer Communications, Vol. 23, 2000, pp. 462–475.

[36] Madan B. B., S. Dharmaraja, K. S. Trivedi, "Combined Guard Channel and Mobile Assisted Handoff in Cellular Networks", IEEE Transactions on Vehicular Technology, Vol. 57, No. 1, 2008, pp. 502–510.

[37] Choi S., K. G. Shin, "Adaptive Bandwidth Reservation and Admission Control in QoS-sensitive Cellular Networks", IEEE Transactions on Parallel and Distributed Systems, Vol. 13, 2002, pp. 882–897.

[38] Pencheva E., I. Atanasov, "Adaptive Load Control for Parlay X Gateway", International Journal on Information Technology and Security, No. 2, 2010, pp. 17–32.

[39] Epstein B., M. Schwartz, "Reservation Strategies for Multimedia Traffic in a Wireless Environment", IEEE Vehicular Technology Conference, Conf. Proceedings, 1995, pp. 165–169.

[40] Yin L., Z. Zhang, Y.-B. Lin, "Performance Analysis of Dual-threshold Reservation Scheme for Voice/Data Integrated Mobile Wireless Networks", IEEE WCNC, Conference Proceedings, 2000, pp. 258–262.

[41] Li B. et al., "Call Admission Control for Voice/Data Integrated Cellular Networks: Performance Analysis and Comparative Study", IEEE JSAC, Vol. 22, No. 4, 2004, pp. 706–717.

[42] Alouini M. S., A. J. Goldsmith, "Adaptive Modulation over Nakagami Fading Channels", Journal on Wireless Communications, Vol. 13, No. 1–2, 2000, pp. 119–143.

[43] Wang H., V. Iversen, "Erlang Capacity of Multi-class TDMA Systems with Adaptive Modulation and Coding", IEEE International Conference on Communications (ICC'08), Conference Proceedings, May 2008, pp. 115–119.

[44] Qingwen L., Z. Shengli, B. Georgious, "Queuing with Adaptive Modulation and Coding over Wireless Links: Cross-layer Analysis and Design", IEEE Transactions on Wireless Communications, Vol. 4, No. 3, May 2005, pp.1142–1153.

[45] Tarhini C., T. Chahed, "System Capacity in OFDMA-based WiMAX", ICSNC'06, Conference Proceedings, Vol. 4, No. 3, 2006, pp.70–74.

[46] Perez D. L. et al., "Intracell Handover for Interference and Handover Mitigation in OFDMA Two-Tier Macrocell-Femtocell Networks", EURASIP Journal on Wireless Communications and Networking, Vol. 2010, 2010, pp. 1–15.

[47] Jabbari B., "Teletraffic Aspects of Evolving and Next-generation Wireless Communication Networks", IEEE Personal Communications, 1996, pp. 4–9.

[48] Guerin R. A., "Channel Occupancy Time Distribution in a Cel-lular Radio System", IEEE Transactions on Vehicular Technology, Vol. 35, No. 3, 1987, pp. 89–99.

[49] Herzog U., L. Woo, K. M. Chandy, "Solution of Queuing Prob-lems by a Recursive Technique", IBM Journal on Research and Development, Vol. 19, 1975, pp. 295–300.

[50] Xue D., X. Wang, "Adoption of Cognitive Radio Scheme to Class-based Call Admission Control", IEEE ICC'2009, Conference Proceedings, 2009, pp. 1–7.

[51] 21 Fodor G., M. Telek, "A Recursive Formula to Calculate the Steady State of CDMA Networks", 19th International Teletraffic Congress (ITC), 2005, pp. 1285–1294.

[52] Kassev K., B. Tsankov, "Intra-cell Handover in OFDMA-based Wireless Access Networks", ICEST'2010, Conference Proceedings, Vol. 1, 2010, pp. 43–46.

5

Enhancing Positioning Accuracy via Heterogeneous Wireless Networks

A. Zvikhachevskaya
Accunostics Ltd. Company, UK

D. Rodionov
Rinicom Ltd Company, UK

L. Mihaylova
University of Sheffield, UK
E-mail: Lyudmila S Mihaylova <l.s.mihaylova@sheffield.ac.uk>

G. Markarian
Lancaster University, UK

Introduction

In emergency situations or at risk the mobile phone is the device which is most likely to be used for the emergency call. Most mobile phones have connection to the Global Positioning System (GPS) which allows accurate outdoor positioning. In indoor environment or when surrounded by high buildings the GPS is not reliable for the accurate positioning, due to multipath or signal blocking. Similar problems exist in terrestrial-based networks in multipath environment.

Alternative solutions for positioning are, wireless local area network technologies (such as WiFi), Worldwide Interoperability for Microwave Access (WiMAX) and Long Term Evolution (LTE). These technologies provide flexibility in accessing the network, high data rates and a single standard approach that does not rely on vendor specific solutions.

This chapter presents methods and solutions that make possible to overcome drawbacks of the GPS by utilizing 4G and 3G technologies for

Vladimir Poulkov and Ramjee Prasad (Eds.), Resource Management in Future Internet, 111–142.

Localisation and Positioning (L&P). A survey of the available wireless networks technologies is given and their signal measurements for positioning services. A novel method for performing positioning services on mobile devices based on coexistence of WiFi, WiMAX, LTE technologies is proposed. In addition, a solution that incorporates new methods using Time of Arrival (TOA) measurements, an Adaptive Modulation and Coding method (AMC) is suggested for the L&P services.

5.1 Positioning Methods and Technologies

Innovations in the area of communication systems and wireless technologies triggered the progress in L&P systems and applications [9, 10]. Location is a kind of situational information [10] and can be in use in many context aware applications such as monitoring, tracking, etc.

Some of the most commonly used wireless systems [10] are:

1. Wireless Local Area Networks (WLANs) such as WiFi, which are mostly used for indoor positioning;
2. Wide Area Networks (WANs) such as cellular communication networks (e.g. GSM and WiMAX);
3. The Global Positioning System (GPS);
4. Radio Frequency Identification (RFID) technologies which rely on remotely storing and retrieving data using tags (transmitters) and readers. Multiple RFID tags are used for the accurate object detection and it is based on the principle of measurements of the RF signal strength between a reader and a tag. A comparison of this measurement with the one in the database reflects the spatial information between them.

The purpose of this section is to provide a comprehensive survey of the technologies that support L&P applications and algorithms.

5.1.1 Radio Frequency Based Positioning Technologies

There are various technologies that allow collection of measurements for the location and position estimation. A classification of the main technologies is given on Figure 5.1 These are radio frequency (RF) based and non-RF technologies (e.g. Infrared, ultrasound, laser). The second group is out of the scope of this chapter.

Wireless networks utilised for localisation can be classified based on the area of coverage. A review of the most popular available technologies for the

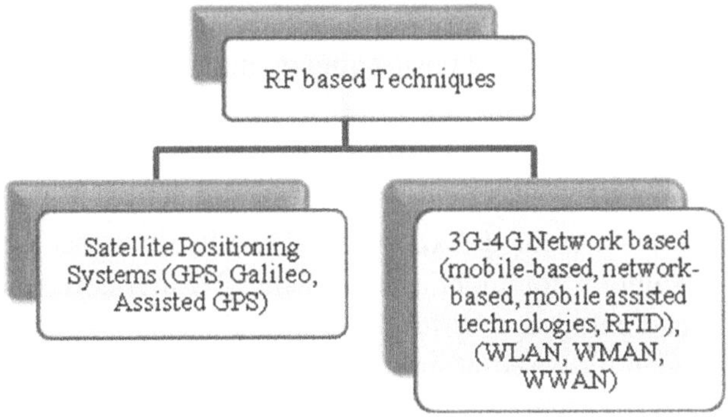

Figure 5.1 Wireless Technologies for Positioning Applications.

Table 5.1 Wireless PAN Technologies

Parameters	Bluetooth (IEEE802.15.1)	UWB (IEEE802.15.3)	ZigBee (IEEE802.15.4)
Applications	Computer to Computer, Computer with other Digital Device, Computer and Accessory devices.	Multimedia Content Transfer, High-resolution radar, Wireless Sensor Network, etc.	Home Control, Building Automation, Medical monitoring, etc.
Frequency Band	2.4–2.8GHz	3.1–10.6GHz	868MHz, 902–928MHz, 2.4–2.48GHz,
Range	~10 meters	~10 meters	~100 meters
Modulation	GFSK, 2PSK, DQSP, 8PSK	OPSK, BPSK	BPSK, OPSK

object location is presented further, covering personal, local, metropolitan and wide area wireless networks (PAN, LAN, MAN, WAN).

5.1.2 Personal Area Network

Within PAN networks Bluetooth, Ultra Wide Band (UWB) and ZigBee technologies are the most commonly used for localisation.

Bluetooth

Bluetooth has found its application in the field of the localisation and positioning. Two are the most popular positioning methods which are based on IEEE 803.15.1 [28] technology:

- Measuring the received signal power level;
- Finding the cell identity based positioning by mapping known Bluetooth devices addresses to location information.

In [28], a system for delivering permission-based location-aware mobile advertisements to mobile phones using Bluetooth and Wireless Application Push (WAP) Protocol is presented. Experimental results show that the system provides a viable solution for mobile advertising. Furthermore, in [29], a Bluetooth-based positioning system for museum guidance is described. The accuracy of the all identity-based positioning techniques can be improved by placing several fixed Bluetooth devices with different reachable distances into a given location and excluding paths that are not physically possible.

Ultra wide band

Examples of UWB Positioning systems are:

- In [30], an UWB location system is presented which employs relative location principles to enhance the performance in wireless networks. The pear-to-pear range measurements between the device and neighbours are used.
- In [31], methods based on ray tracing are developed to reduce the ranging error resulting from the propagation delay. Considerable accuracy improvement is observed in the map-aided positioning.
- Furthermore, in [32], it is shown that the use of power control improves the robustness of location-estimates and provides higher localisation accuracy. The localisation accuracy of a mobile device is determined by connectivity with reference nodes, range-estimate variances and geometry of mobile nodes.
- Yu et al. in [37] investigate the performance of different position estimation methods by using the time-of-arrival of ultra wideband signals. The low cost and low system complexity is successfully tested in a ski field where skiers are tracked and localised.

ZigBee

There are many techniques for location and positioning which are based on ZigBee:

- ZigBee can be used as an indoor positioning system for object tracking. For example in [34], is shown that wireless nodes equipped with sensors can provide unique identifiers, effectiveness and flexibility.

- Another ZigBee location estimation system [35] is tested in home networking environment. The maximum likelihood location estimation (MLLE) algorithm is presented. It is proposed to use cluster free topology for the Received Signal Strength (RSS) measurements. The algorithm provides a sufficient accuracy to support services for the home environment.
- In [36], an application of the fingerprinting method to the location estimation in ZigBee networks is presented. Experimental results demonstrate the validity and 70% accuracy with the tolerance of 0.5 meters.

5.1.3 Wireless Local Area Network (WLAN)

Principles of work for WLAN localisation is based on the signal strength between an object and beacons [11]. Advantages of the WLAN localisation are as follows:

- economical, light weight with the low power consumption device;
- already well deployed network infrastructure (WLAN access points (APs)).

The main disadvantage of the WLAN localisation is that it requires a wired connection with a computer. Hence it is not very portable and suitable for the indoor environments.

Localisation and Positioning by using WLAN APs is an active research field and there are numerous results. Here only few of them are surveyed. In [12], a wireless LAN-based indoor position technology is presented. WLAN signals are used to create a model-based signal distribution training scheme. This scheme allows finding an optimal solution with respect to accuracy of the signal distribution and the training workload. An algorithm for object tracking is presented to utilise the knowledge of the topology and assist the positioning process. In [12] experimental results from a WLAN-based indoor positioning system are given. In 90% of the results the accuracy varies from 2 to 5 meters depending on the environment and speed of the mobile.

Another example of the localisation based on IEEE 802.11 technology is presented in [13] and [14]. In [13], a localisation system relying on the IEEE 802.11b wireless network and a multilateration algorithm is developed in conjunction with a quadratic polynomial representation of the distance-signal strength relation. Similarly, in [14], a wireless localisation device that uses Bayesian networks to infer the location of objects covered by IEEE 802.11 wireless network is introduced and tested. In [15] the system named "RADAR" is proposed and a nearest-neighbour method is used for the position estimation. A signal propagation model is proposed to describe the signal propagation. In [16] a probabilistic approach to estimate user location is proposed. In [17] a positioning system based on a neural network model is presented.

In [18], localisation results based on a PCMCIA wireless Ethernet card are presented. The achieved estimated location accuracy is less than 1 meter. In [19], the WLAN is used for localisation of mobile robots. It is shown that given an adequate number of beacons and a signal strength map, robots can be localised with an error of 0.5 m. In [20] a table-based method of position determination is proposed. In [21] a tracking-assistant positioning system is introduced. Tracking techniques are also used by the Ekahau [22] system.

5.1.4 Wireless Metropolitan Area Network (WMAN)

The IEEE 802.16/WiMAX technology provides broadband wireless services a with wide service coverage, high data throughput and mobility support (IEEE 802.16e). Localisation and positioning by using WiMAX networks is an active research field. In [26], a linear WiMAX based station (BS) serves as a roadside unit and a WiMAX mobile station (MS) installed in a vehicle serves as an on board model to study the performance of the location update with or without anchor paging controller (APC) relocation.

The topology of WiMAX networks is similar to those of the GSM. Both of them use BSs to establish a wireless connection with subscriber stations (GSM terminal or WiMAX enabled computer for example). Almost the same quantities can be measured using both networks. In WiMAX networks the following signals can be used for localisation:

- The Timing Adjust (TA): This concept is similar to Timing Advance (TA) or Time of Arrival (TOA) concept in GSM networks [15].
- The Time Difference of Timing Adjust (TDOTA): This concept is similar to the Time Difference of Arrival (TDOA) concept in GSM networks. The idea of this measurement is to compare more than one TA values measured to different base stations to eliminate the measurement error caused by the terminal clock synchronisation.
- The Angle of Arrival (AOA): WiMAX uses directional antennas which allow the determination of the azimuth of a terminal seen by a certain BS. The antennas used in Pre-WiMAX network in Brussels provide this information as sectors (60, 90 and 120 degrees). WiMAX networks started to use advanced antenna arrays where beam forming allows

rotating narrow beams. The narrow antenna patterns will increase the accuracy of the measured terminal azimuth.

- The Base Station Identifier (BSID): This concept is the same as Cell-ID in GSM networks. The position of a terminal can be determined depending on the serving base station coordinates. This value can be obtained –by the terminal- by obtaining the serving base station MAC address which is broadcasted over the control channel.
- The Received Signal Strength Index (RSSI): WiMAX terminals can measure the received signal power broadcasted by a BS although extra software is needed. The RSSI measurement gives information about the distance between the terminal and the corresponding BS. The RSSI values depend on the operating environment and a path loss model has to be developed for a respective propagation environment.
- The SCORE values: The current standard WiMAX terminals measure the SCORE values of the available BSs. The SCORE values are related directly to the RSSI values. However, they can be considered, with some approximations, as rough RSSI measurements.

In addition, the support of short-range communications among the terminals (mesh networks) in WiMAX networks is proposed in [15]. The rationale for introducing short range communications is mainly due to three reasons:

1) The need to extend the coverage to places not covered by a BS.
2) Support peer-to-peer (P2P) high-speed wireless links between the terminals.
3) The need to enhance the communication between a terminal and the base station by fostering cooperative communication protocols among spatially proximate devices.

The accuracy of the location estimation can be enhanced by utilising the additional information gained from measuring the relative distances between the terminals. The support of short-range communications is very attractive, but the practical implementation is complicated and has a lot of implications. Therefore, the use of mesh networks could be avoided or be limited to security and emergency cases.

For example, a police car (or an ambulance) can establish a direct connection to other cars; or in case of being outside the coverage area of the wireless network, a connection can be established to the main network backbone by using the available modems in its range.

Therefore, WiMAX networks have all the resources to locate their subscribers without relying on any external system. The most attractive resources for localisation are the ones that are easy to obtain and already available during the normal terminal operation such as RSSI-based values.

5.1.5 Satellite Positioning Systems

The GPS is a space-based global navigation satellite system that provides reliable location and time information in all weather and at all times and anywhere on or near the Earth when and where there is an unobstructed line of sight to four or more GPS satellites [38].

5.1.6 Examples of Positioning in Various Environments

Elderly people health monitoring and tracking

The main purpose of health control services is to perform continuous monitoring of vital parameters such as heart rate, blood pressure, temperature etc. for every patient. The service handles data from sensors and informs operators in case of dangerous shifts of these parameters from normal values. Thus, such systems require build-in location and positioning service, which helps to find the person in case of emergency.

Problems and limitations of tracking patients in a hospital or a nurse house in cases of emergency occur due to various reasons:

- Medical equipment cannot operate via Wi-Fi and Bluetooth as they use only high frequency bands. Thus wireless communication must be based on low-frequency protocols;
- Another problem is due to energy saving requirements, as the battery charge must be sufficient for working more than a year or even more. To overcome this issue, wireless sensors should send data periodically with a big interval of 5–10 seconds. Thus, the system must estimate position after the last received signal from the tracked sensor. Also, if sensors are broken or the battery is discharged, the system must predict the position in future time after the last signal is received.

Additional benefits from the positioning system is the possibility of continuous monitoring which allows storing typical routes and locations of the patients and estimating abnormal behavior. For example, when the patient is staying or a long time in the corridor, the system should report it to the operator and, for instance, show video image from the corridor camera.

Numerous solutions to the location and positioning problem are proposed but it is still an unresolved problem.

Passenger tracking in an airport

There are a lot of problems when one or several passengers are late for boarding. There are consequences of a delayed flight as:

- all passengers could be late for connection flight,
- this could cause fines to the airline service providers;
- this leads to additional complications for the airport services.

Video surveillance

Video surveillance systems which include dome cameras, a remote control operator panel, a video storing server, and a coordinate's estimation server can also serve for location finding purposes. All cameras are provided with compass and GPS and they are able to determine the direction and position of the tracked objects.

However, some parts of the monitored territory can be unreachable for the video cameras. Also there are occasions due to with out of order cameras. Then position prediction becomes the only possible solution for positioning services.

5.2 Review Methods on Cooperative Mobile Positioning

The main research focus on the mobile positioning systems is on increasing the accuracy of already existing techniques or to employ new evolving wireless technologies for the positioning services.

However, cooperative mobile positioning in hybrid networks is a relatively novel area which usually includes cooperation minimum two different types of wireless positioning technologies, e.g. WiFi & UWB [43], WiMAX & WiFi [27] and sensor networks [42–43]. Furthermore, in [40], positioning methods which are based on RFID, Bluetooth and WLAN technologies are presented. Here, the focus is made on expected coexistence of common RF technologies such as passive UHF RFID tags/readers, RSSI measurements from Bluetooth and the TOA information from the IEEE 802.11. The accuracy achieved fluctuates from 0.5 to 4 meters.

Other advances are related with hybrid data fusion techniques, which fuse for instance TOA with AOA, TDOA with RSS measurements, etc. For example, in [39], a hybrid scheme is developed combining TOA with AOA measurements for mobile localisation in cellular communication systems. It is proposed in [39] to integrate the available measurements: the TOA from the BSs with the AOA information at serving BS. As a result, the proposed methods provide better accuracy compared with the Taylor Series Algorithm

(TSA) and the Hybrid Lines of Positioning Algorithm (HLOP). Another example of cooperative positioning techniques for mobile localisation in 4G networks is presented in [41]. This work explains the concept of the cooperative localisation by utilising the additional information obtained from the short-range links (such as WiFi) to enhance the location estimation accuracy in cellular networks. Simulations with a hybrid WiMAX and WiFi system are carried out where TDOA and RSS measurements are combined via data fusion techniques.

Data Fusion algorithms detect signal measurements from the available in the range BSs to calculate an estimate of MS location. Consider the MS and at least three BSs in Cartesian (x and y) coordinates, whilst the z coordinate is ignored. Methods such as positioning using TOA with MIMO features, beamforming and AMC are presented further.

5.2.1 Positioning via Time of Arrival using MIMO Features

The TOA data fusion technique is based on measuring the propagation delay of the radio signal (as opposed to a data packet) between a transmitter (tag) and one or more receivers (readers). The propagation delay (t_i - t_0), is the time lag of the departure of a signal from a source station to a destination station.

A concept of the TOA method [32] is in multiplying the propagation time (t_i - t_0) by the propagation speed of the signal, the propagation delay can be converted into a distance between the transmitter (tag) and the receiver (reader).

Disadvantages of the TOA method are:

- the clocks of the tag and the reader must be *synchronised* in order to have confidence in the measurement of the elapsed time;
- the clock synchronisation system must be developed which has high *costs* in terms of development time and effort;
- three or more readers are required, which also adds to the *cost* and *complexity* of the system.

The distance between the MS and BSs are calculated then using a trilateration algorithm [33] and the conventional Linear Least Squares (LLS) and non-LLS estimation [36] algorithms. The case with two BSs is illustrated on Figure 5.2. The following equations are derived to calculate the MS position with LLS. Note that in order to simplify the calculations, the equations are formulated so that the centers of the spheres are on the z = 0 plane. Also the formulation is such that one center is at the origin.

$$r_1^2 = x_m^2 + y_m^2, \tag{1}$$

$$r_3^2 = (x_3 - x_m)^2 + (y_3 - y_m)^2, \tag{2}$$

Then subtracting (1) from (2) and (3) from (1) gives

$$r_2^2 - r_1^2 = x_2^2 - 2\,x_2\,x_m + y_2^2 - 2y_2\,y_m, \tag{3}$$

$$r_3^2 - r_1^2 = x_3^2 - 2x_3\,x_m + \,y_3^2 - 2y_3\,y_m, \tag{4}$$

The above equations can be written as

$$\begin{bmatrix} x_2 & y_2 \\ x_3 & y_3 \end{bmatrix} \begin{bmatrix} x_m \\ y_m \end{bmatrix} = \frac{1}{2} \begin{bmatrix} K_2^2 & - & r_2^2 & + & r_1^2 \\ K_3^2 & - & r_3^2 & + & r_1^2 \end{bmatrix}, \tag{5}$$

where $K_i^2 + x_i^2 + y_i^2$,, the equation can be rewritten

$$H_x = b,$$

$$H = \begin{bmatrix} x_2 & y_2 \\ x_3 & y_3 \end{bmatrix}; x = \begin{bmatrix} x_m \\ y_m \end{bmatrix}; b = \tfrac{1}{2} \begin{bmatrix} K_2^2 & - & r_2^2 & + & r_1^2 \\ K_3^2 & - & r_3^2 & + & r_1^2 \end{bmatrix} \tag{6}$$

where

Note the equations so far are for three measured distances only. If more measurements are present they can be added to the equation as well.

$$H = \begin{bmatrix} x_2 & y_2 \\ x_3 & y_3 \\ \vdots & \vdots \end{bmatrix}; b = \frac{1}{2} \begin{bmatrix} K_2^2 & - & r_1^2 & + & r_1^2 \\ K_3^2 & - & r_3^2 & + & r_1^2 \\ & & \vdots & & \end{bmatrix} \tag{7}$$

So the LLS solution is

$$x = (H^T\,H)^{-1} H^T.$$

The MIMO technology is a wireless technology that uses multiple transmitters and receivers to transfer more than one signal. It takes advantage of a radio-wave phenomenon called multipath where a signal reaches the receiving antenna multiple times via different paths and thus different angles and times [34]. Furthermore, the MIMO technology controls the multipath by using multiple, "smart" transmitters and receivers with an added "spatial" dimension to dramatically improve performance with the increase of the range.

Taking position via TOA by utilising MIMO features was proposed in [32]. We decided to study further positioning via TOA utilising MIMO features by applying it to the proposed heterogeneous wireless network scenario and the cooperative scheme. The MIMO antenna is considered as diversity antenna, so that only TOA measurements are taken into account.

It is possible to study the performance when the conventional 1 x 1 systems use single antennas at both ends and to compare the results with the MIMO case. Hence, using MIMO, more signals are detected by MS and hence the location estimation accuracy is high. The N_t *transmission antennas send* different signals simultaneously over minimum of N_t x N_r transmission paths and each of the N_r *received antennas* provide signals which represent a combination of the N_t transmitted signals and distorting noise.

Hence, as it is shown in [32] and proved by our implementation, MIMO is not only improving the capacity of a wireless link but also can be used to advance the accuracy of the positioning system.

Hence, considering the MIMO TOA based method with M BSs estimates [32], the mobile position estimates are:

$$x_{MIMO} = \arg\min_x \sum_{i=1}^{M} \sum_{i=1}^{P} \sum_{j=1}^{n} \delta_{i,n}^{1} - ||x - x_i||^2, \quad (8)$$

where x is the true position of the MS, x_i and δ, denote the coordinates of i-th BS and the range measurement between the i-th BS and the MS with δ^l where l - pilot (preamble) signal, respectively, I = 1, 2,...,M and n = 1,2,..., N_t–N_r.

In (11) ||.|| denotes the Euclidean norm and |.| the absolute value.

5.2.2 Positioning Using Adaptive Modulation and Coding Data Fusion

Adaptive Modulation and Coding allows wireless systems (such as 3G, WiFi, WiMAX, etc.) to adjust the signal modulation scheme depending on the signal-to-noise (SNR) condition of the radio link. The idea behind the AMC is to dynamically adapt the modulation and coding scheme to the channel condition so as to achieve highest spectral efficiency at all times. However, this causes the signal level to be almost the same throughout a BS coverage area, so that the signal level measurement cannot be used to estimate the location of mobile users. In addition, wireless technologies use power control to adjust the signal quality based on the SNR. Nonetheless, by taking the information from both the physical layer (the type of modulation scheme used and power control reading) and MAC layer at the WiMAX BS, the data can be used to determine the MS's location as shown in Figure 5.3. In each region, users have the same modulation and coding scheme (MCS). In order to determine the area covered by each modulation scheme, the maximal distance, R_i between the BS and MS must be calculated first using a corresponding modulation. This distance is determined using the maximal SNR a user should receive without data loss. Different values of the received SNR for different MCS have been calculated and are shown in Tables 5.1, 5.2 and 5.3. Then, R_i can be calculated using this information. Suburban and urban areas are considered as types of communication environment [35]:

$$PL_i\,[dB] = -10\log\left[\frac{\lambda^2 G_t G_r}{(4\pi R_i)}\right], \tag{9}$$

where λ is the wavelength, G_t and G_r is the transmitter and receiver antenna gain, respectively and R_i is the distance between the transmitter and the receiver. Equation (12) can be written in the form:

$$PL_i = P_i[dB] - SNR[dB] - N[dB] \tag{10}$$

where P_t is the transmitter power and N is the thermal noise which is given by:

$$N[dB] = 10\ \log(\tau TW), \tag{11}$$

where $\tau = 1.38 .10^{-23}\,JK - 1$ is the Boltzman constant, T is the temperature in Kelvin (T = 290) and W is the transmission bandwidth in Hz.

Using the above equations, we can calculate the relationship between the distance and the SNR as follows:

$$R_i = \frac{\lambda \cdot 10^{\frac{P_c + 10\log(G_t) + 10\log(G_t) - SNR - N}{20}}}{4\pi} \tag{12}$$

After calculating formula (15) for each distance between a BS to a MS, estimates are obtained for the MS coordinates, using linear least squares and non-linear least squares algorithms. The biggest advantage of this method for location and positioning services is that it can be used to estimate how far away the MS is from the BS using just one BS. Hence, the higher the number of BSs is, the more accurate the location estimate is.

The free space loss model is not applicable, because there are multiple factors that can influence the propagation of the electromagnetic waves.

Hence, to create a more realistic scenario the Cost 231-Hata model [37] is used. It is based on a series of measurements made in 1968, in and around Tokyo in the band between 150 MHz and 1500 GHz. The results are used in series of graphs for path loss predictions. It allows considering urban environment and city.

The simulation set-up and network architecture are described in Section III and shown in Figure 5.1 Computer simulations are performed to demonstrate an accuracy of the proposed location scheme. We consider a hexagonal cell (where the serving BS resides) as shown in Figure 5.1 Each cell has a radius corresponding to their standard possibilities in meter. 1000 independent trials are performed for each simulation. Regarding the NLOS effects in the simulations, the Circular Disk of Scatterers Model (CDSM) [32] is performed to assess the performance of all methods. The CDSM assumes that there is a disk of scatterers around the MS and that signals between the MS and the BSs undergo a single reflection at a scatterer. The measured positions via DOA-based, TOA and AMC at the serving BS (e.g. BS1) would be the angle, time of arrival or SNR between the BS and a scatterer. The measured ranges are the sum of the distances between the BS and the scatterer and between the MS and the scatterer. If BS1 is the serving BS, its measurements should be more accurate. The radius of the scatterers for BS1 is e.g. 250m for the WIMAX and the other BSs were taken from 20m to 250m, for the WiFi and LTE wireless technologies, respectively. As the radius of the scatterers increases, the average magnitudes of the NLOS time and angle errors increase and lead to less accurate location estimation.

System level settings

System level simulation parameters for the WiMAX, WiFi and 3GPP LTE networks were set according to standard requirements. Each of the BS to MS link is unique and corresponds to the relative standard. It is assumed that MS is compatible with all three technologies. The BS to MS links is set as following.

BS of the IEEE 802.16 (WiMAX) network - MS link

IEEE 802.16 (WiMAX) is modeled based on the carrier frequency and the system bandwidth 3.4GHz and 20MHz, respectively. The Physical layer (PHY) adopted from the IEEE 802.16 standard which is based on the Orthogonal Frequency Division Multiplexing (OFDM). Table I summarises the specific modulation scheme parameters considered in the simulations for the WiMAX technology.

BS of the IEEE 802.11g (WiFi) - MS link

In Table 5.1 the MCS on the licensed band IEEE 802.11 (WiFi) technology (carrier frequency and system bandwidth 2.4GHz and 20MHz, respectively) is shown. The Physical layer (PHY) adopted from the IEEE 802.11g / 2.4 GHz standard which is based on OFDM. Table 5.2 summarises the specific parameters considered in the simulations.

BS of the 3GPP LTE - MS link

The use the MCS for the 3GPP LTE technology is presented in Table 5.1. The carrier frequency and system bandwidth are 20MHz and 15 kHz, respectively. The Physical layer (PHY) adopted from the 3GPP LTE standard employs OFDM for the downlink data transmission and single carrier frequency division multiple accesses (SC-FDMA).

Table 5.2 Received SNR assumptions for IEEE 802.16E, IEEE 802.11G and 3GPP LTE

Modulation	Coding Rate			Receiver SNR (dB)		
	IEEE 802.16e	IEEE 802.11g	3GPP LTE	IEEE 802.16e	IEEE 802.11g	3GPP LTE
BPSK	1/2	1/11	-	3.0	4.0	-
QPSK	1/2	1/11	1/2	6.0	7.0	3.8
	3/4	3/4	3/4	8.5	15.0	5.9
16-QAM	1/2	1/2	1/2	11.5	17.0	10.0
	3/4	3/4	3/4	15.0	21.0	12.3
64-QAM	2/3	2/3	2/3	19.0	25.0	17.0
	3/4	3/4	3/4	21.0	26.0	18.0

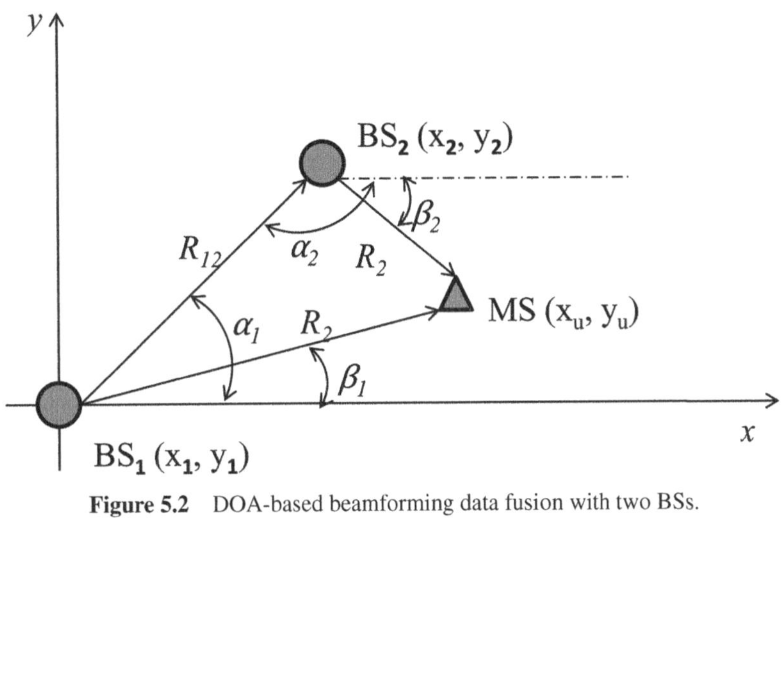

Figure 5.2 DOA-based beamforming data fusion with two BSs.

5.2.3 Positioning Using Beamforming Features

Beamforming is one of WiMAX features commonly used to boost both capacity and coverage. It will be employed here to improve the location estimation accuracy. Beamforming is a method used to create the radiation pattern of an antenna array. Beamforming utilises multiple antenna elements, or arrays, as is the case with diversity and MIMO techniques. There are two prevalent beamforming techniques namely Direction of Arrival (DOA)-based beamforming and eigenbeamforming. They differ from one another regarding the direction toward which energy is focused [35].

DOA-based beamforming is based on physical direction, while eigen-beamforming (also known as intelligent beamforming) is based on the mathematical direction. In this chapter, our focus will be the first technique by using parameter measurements of TOA and DOA for MIMO system. The DOA of MS signals at a BS can be obtained by antenna arrays. The DOA of the MS signal can be calculated by measuring the phase difference between the antenna array elements or by measuring the power spectral density across the antenna array.

More generally, assume *M* BSs estimate the DOA of the MS signal, and the goal is to combine these measurements to estimate the MS location. As indicated in Figure 5.2, let β_1 and β_2 denote the AOA of the MS signal at BS_1 and BS_2, respectively.

Then we have

$$\begin{bmatrix} x_u \\ y_u \end{bmatrix} = \begin{bmatrix} x_1 \\ y_1 \end{bmatrix} + \begin{bmatrix} R_1 \cdot \cos\ \beta_1 \\ R_1 \cdot \sin\ \beta_1 \end{bmatrix}, \tag{13}$$

and

$$\begin{bmatrix} x_u \\ y_u \end{bmatrix} = \begin{bmatrix} x_2 \\ y_2 \end{bmatrix} + \begin{bmatrix} R_2 \cdot \cos\ \beta_2 \\ R_2 \cdot \sin\ \beta_2 \end{bmatrix}, \tag{14}$$

where

$$\begin{aligned} R_1 &= \sqrt{R_1^2 + R_{12}^2 + 2R_1R_{12}\cos(\alpha_1 - \beta_1)} = f(\alpha_2, \beta_2, R_2, R_{12}) \\ R_2 &= \sqrt{R_1^2 + R_{12}^2 + 2R_1R_{12}\cos(\alpha_1 - \beta_1)} = f(\alpha_1, \beta_1, R_1, R_{12}) \end{aligned} \tag{15}$$

Since α_1, α_2, β_1, β_2, R_{12} is known, we simply denote R_1 as a function of R_2 as $R_1 = f_1(R_2)$, and so on R_2 as a function of R_1 as $R_2 = f_2(R_1)$.

Likewise, for any other BS_1:

$$\begin{bmatrix} x_u \\ y_u \end{bmatrix} = \begin{bmatrix} x_i \\ y_i \end{bmatrix} + \begin{bmatrix} R_i & \cdot & \cos & \beta_i \\ R_i & \cdot & \sin & \beta_i \end{bmatrix}, \tag{16}$$

In a case there are more than two BSs, a LLS formulation can be obtained by collecting the relations in above equation into a single equation as

$$\vec{b} = A\vec{\theta}, \tag{17}$$

where

$$\vec{b} = \begin{bmatrix} x_1 + R_1 \cos \beta_1 \\ y_1 + R_1 \sin \beta_1 \\ -------- \\ x_2 + R_2 \cos \beta_2 \\ y_2 + R_2 \sin \beta_2 \\ -------- \\ - \\ - \\ - \\ -------- \\ x_i + R_i \cos \beta_i \\ y_i + R_i \sin \beta_i \end{bmatrix}, A = \begin{bmatrix} 10 \\ 01 \\ -- \\ 10 \\ 01 \\ -- \\ \cdots \\ \cdots \\ -- \\ 01 \\ 01 \end{bmatrix}, \vec{\theta} = \begin{bmatrix} x_u \\ y_u \end{bmatrix}, \tag{18}$$

The least square solution for θ is then

$$\vec{\theta} - (A^T A)^{-1} A^T \vec{b} \tag{19}$$

5.2.4 Hybrid Data Fusion Techniques

Combinations of two or more different schemes are used to enhance the location accuracy. In this section hybrid data fusion schemes are described. The main idea is to obtain the position estimates by one of the algorithm and then by utilising additional signal measurements such SNR or beamforming to reduce an estimated error. Figure 5.5 illustrates the performance of the developed algorithm.

5.2.4.1 Hybrid TOA with MIMO features and AMC data fusion

In TOA and AMC methods three or more BSs are involved in the MS positioning process. The MS is not located at the same distance from each

BS (within WiMAX or Heterogeneous network) and conditions of the radio link (e.g. path loss, the interference, the sensitivity of the receiver, etc.) can vary. Hence, link adaptation, or AMC in wireless systems is used to match modulation scheme and other signal parameters to the conditions of the link. In this case, an alternate data fusion procedure is used to obtain AMC location estimates and combine them with TOA estimated considering features of antenna (Figure 5.5). In real life scenarios, the accuracy of TOA and AMC estimates is usually a function of an environment and type of antenna.

The following is a two-layer hybrid least square algorithm. Assume n BSs estimate the AMC and TOA of the MS in the (21). Likewise, from the formula (15), the *(x,y)* estimates using only SNR measurements is given by (22).

$$(\hat{x},\hat{y})_{MIMO\_TOA} = \arg\min_{(x,y)} \sum_{i=1}^{M}\sum_{i=1}^{P}\sum_{j=1}^{n} \delta_{i,n}^{1} - \left\|(x,y)-(x_1,y_i)\right\|^2, \tag{20}$$

$$(\hat{x},\hat{y})_{MIMO\_AMC} = \arg\min_{(x,y)} \sum_{i=1}^{M}\sum_{i=1}^{P}\sum_{j=1}^{n} \delta_{i,n}^{1} - \left\|(x,y)-(x_1,y_i)^2\right\|, \tag{21}$$

where *(x,y)* is the true position of MS, (x_i, y_i) and δ, denote the coordinates of *i-th* BS and the range measurement between the i-th BS and the MS with *l*-th pilot (preamble) signal, respectively, i =1,2,...,M and n = 1,2,..., $N_t \times N_r$. N_t *(the transmit antenna) sends* different signals simultaneously over minimum of N_t*N_r transmission paths and each of those *number of received antenna* (N_r) received signals is a combination of all the N_t transmitted signals and distorting noise.

The final location estimate is calculated as a linear combination of the two estimates:

$$(\hat{x},\hat{y})_{AMC\_TOA} = \eta(\hat{x},\hat{y})_{MIMO\_AMC} + (1-\eta)(\hat{x},\hat{y})_{MIMO\_TOA}, \tag{22}$$

where η is a positive parameter which is chosen depending on the relative accuracy of the TOA and AMC measurements.

5.2.4.2 Hybrid TOA with MIMO Features and DOA-Based beamforming data fusion

In TOA and beamforming positioning methods, two or more BSs are involved in location process. In case of low accuracy of TOA data fusion method, due to the poor channel condition, an alternate data fusion procedure can be used to obtain AOA measurements combined with TOA estimates. For example, in urban areas, beamforming is less accurate than TOA if BS array is surrounded by many scatterers. The hybrid least-square procedure is described below. Assume n BSs estimate the TOA of the MS as presented in (21).

The least square estimate of (x,y) position of the MS using the beamforming data fusion method is given by (24).

$$(\hat{x}, \hat{y})_{DOA} = (A_{DOA}^T A)^{-1} A_{DOA}^T \vec{b}, \tag{23}$$

where

$$\vec{b} = \begin{bmatrix} x_1 + R_1 \cos\beta_1 \\ y_1 + R_1 \sin\beta_1 \\ ------- \\ x_2 + R_2 \cos\beta_2 \\ y_2 + R_2 \sin\beta_2 \\ -------- \\ - \\ - \\ - \\ -------- \\ x_i + R_i \cos\beta_i \\ y_i + R_i \sin\beta_i \end{bmatrix}, A = \begin{bmatrix} 10 \\ 01 \\ -- \\ 10 \\ 01 \\ -- \\ :::::: \\ -- \\ 10 \\ 01 \end{bmatrix}, \tag{24}$$

Finally, the location estimate is calculated depending on the relative accuracy of TOA/MIMO and DOA-based beamforming measurements (26).

$$(\hat{x}, \hat{y})_{DOA\_TOA} = \eta(\hat{x}, \hat{y})_{DOA} + (1 - \eta)(\hat{x}, \hat{y})_{MIMO\_TOA}, \tag{25}$$

where η is a positive parameter which is chosen depending on the relative accuracy of the TOA and DOA-based beamforming measurements.

5.3 Wireless Network Architecture and Positioning Protocol

Positioning in urban environments via dense heterogeneous wireless networks is a more realistic approach then for positioning via WiMAX tools only, meaning that the mobile station (MS) will be able to connect to independent wireless networks, e.g. WiFi, WiMAX and 3GPP LTE, etc.

In Figure 5.3 the studied heterogeneous wireless network architecture is presented. As it can be observed, at least three BSs are in the connectivity range of the MS at the specific time. Hence, the following conditions are necessary:

i) The base stations (BSs) that are in the range of the MS should be synchronised via an adaptive mechanism;

ii) The MS has capabilities to connect to every wireless BS in its range. This is a realistic assumption as all of the chosen standards can be interconnected on the Internet Protocol (IP) layer.

Since there is no central control unit in heterogeneous wireless networks, the following adaptive synchronisation algorithm is proposed for positioning:

1. *Find Available Connections*: In order for the MS to enable the positioning service it is necessary to search all available wireless networks in its range. It is desirable to find more than two BSs as it will guarantee the best accuracy of location estimates.

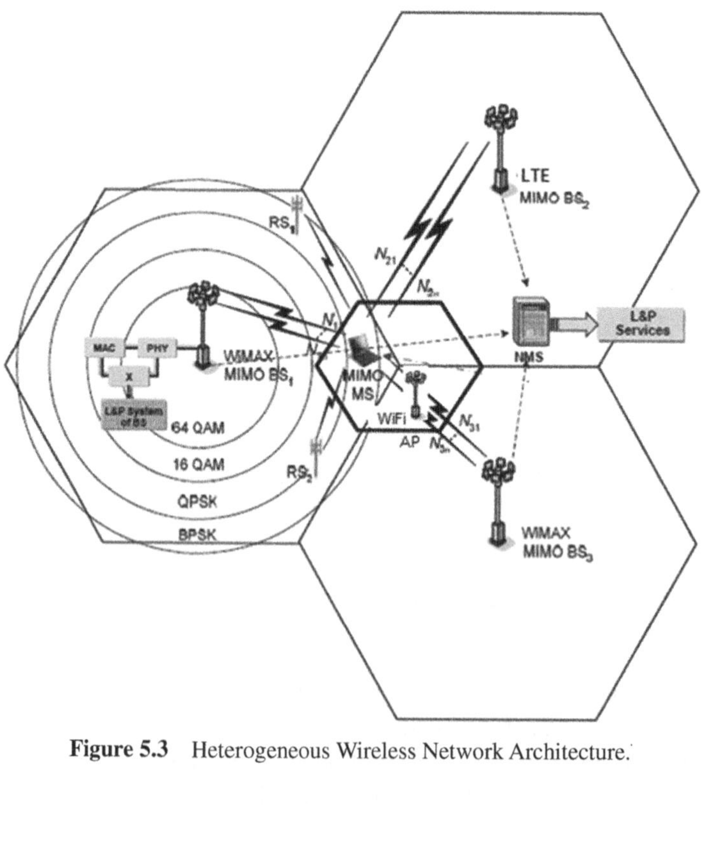

Figure 5.3 Heterogeneous Wireless Network Architecture.

2. *Registration and coordinate request*: Once the "handshaking" is performed it is possible to register the TOA (response time), and send a "coordinate" and modulation scheme requests to all the available in range BSs.
3. *Initial synchronisation*: After registering BS coordinates and corresponding TOA (response times), it is possible to enable the mobile-centric DOA-base (beamforming) calculation algorithm, as all the needed measurements are available.
4. *Data fusion and positioning*: Having TOA, DOA measurements and knowledge of the AMC scheme, MS can calculate location estimates based on the proposed data fusion algorithm, described in Section IV. In the case of the network-centric solution, the MS collects all the available information and redirects it to the Network Management System (NMS presented in Figure 5.1) to perform the positioning task.
5. *Enabling adaptability*: After completing steps 1 to 4 and in order to provide a positioning service of the highest possible accuracy it is essential for the algorithm to adapt to environmental changes. In urban environments, changes can take place due to emergency situations when new dedicated wireless networks are established, due to loss of the previous time-honored connection, changes related to the MS position, and due to new wireless technologies which can appear/disappear from the MS range. Hence, it is proposed to repeat the previous four steps of this algorithm not only in the preset period of time but also taking into account the particularities of the event.

5.4 Findings and Performance Evaluation

This section presents results from the simulation of the WiMAX and heterogeneous wireless networks with the proposed hybrid data fusion methodology. A MIMO communication system is considered in which the effects of the AMC and beamforming are studied.

The AMC positioning method illustrates the effect of coverage area from the BS upon the usage of Modulation Coding Scheme (MCS) on the heterogeneous wireless network. IEEE 802.16 (WiMAX) is modeled based on the carrier frequency and the system bandwidth 3.4GHz and 20MHz, respectively, shown in Table 5.1. In Table 5.1, the MCS on the licensed band IEEE 802.11 (WiFi) technology (carrier frequency and system bandwidth 2.4GHz and 20MHz, is presented.

The effect of the AMC on the 3GPP LTE system, with the carrier frequency and system bandwidth 20MHz and 15kHz, respectively is studied.

The antennas in BS and MS are considered without any gain. We run 500 samples for different SNR values and the results are presented in Figure 5.4 – 5.6. As it is shown, a specific MCS will be used when MS are at a certain distance (formula (5)) from the BS. If a MS is using a particular MCS, it is possible to estimate the distance from the BS. This is achieved by assuming that the selection of the MCS is based solely on the distance from the BS and other factors such as channel conditions do not influence the selection of that particular MCS. Hence, the distance estimates via AMC can be used to provide a hybrid data fusion technique for heterogeneous wireless networks.

Computer simulations are performed to demonstrate the performance of the proposed location scheme. We consider a hexagonal cell (where the serving BS resides) as shown in Figure 5.3. Each cell has a radius corresponding to their standard capabilities in meters. One thousand independent trials are performed for each simulation. Regarding the NLOS effects in the simulations, the CDSM is performed to assess the performance of all methods.

The CDSM assumes that there is a disk of scatterers around the MS and that signals between the MS and the BSs undergo a single reflection at a scatterer. The measured positions via AOA, TOA and AMC at the serving BS (BS1) would be the angle, time of arrival or SNR between the BS and the scatterer. The measured ranges are the sum of the distances between the BS and the scatterer and between the MS and the scatterer. If BS1 is the serving BS, its measurements should be more accurate. The radius of the scatterers for BS1 is e.g. 250m for the WIMAX and for the other BSs were taken from 20 m to 250 m (for the WiFi and LTE wireless technologies). As the radius of the scatterers increases, the average magnitudes of the NLOS time and angle errors increase and this leads to less accurate location estimation. The performance of the developed algorithms was measured by comparing the average root-mean-square (RMS) error (meter). The proposed hybrid TOA/AMC and TOA/AOA positioning schemes still can reduce the RMS errors effectively and accurately estimate the MS location. The sensitivity of the proposed schemes with respect to the NLOS effect is obvious. Figures 5.8–5.15 show cumulative distribution functions (CDFs) of the location error.

In the case of MIMO simulation, we compare the location estimation accuracy at different types of base antenna mode between SISO, MIMO2x2 and 4x4 antennas. The simulated distance error has a total number of 1000 independent data sets and the MS position is obtained by averaging over all the 1000 estimates. In the TOA measurements, the range data are created by calculating the distance from a true MS's position to the known BS positions and the measurement noise and NLOS are added to the true calculated range to get the measured range data. Then the estimate of MS's position can be determined by using the trilateration method based on the LLS and non-linear least square (NLLS) algorithms.

It is observed that with the increase number of antennas at BSs, a better range accuracy can be achieved. Figures 5.7–5.12 show performance results of all localisation based data fusion methods. Results for the SISO, MIMO 2x2 and MIMO 4x4 antenna cases are presented. The accuracy of the SISO system improves when the data fusion scheme is used. Taking into consideration the constraint on ability when power control is employed in wireless cellular systems, to minimize interference the number of BSs involved in positioning should be limited. In order to evaluate the improvement for the various antenna mode configurations, the cumulative probability for the squared value of location error under NLLS algorithm is calculated. Figures 5.8 – 5.13 show the average root mean square error (RMSE) for the TOA positioning data fusion algorithm compared with the developed hybrid TOA-AMC and TOA-DOA algorithms for the IEEE 802.16/WiMAX network.

FCC requirements for the network and handset-based location requirements should be no more than 65% outage with 100(m) and 50(m) location error.

After adding noise, NLOS and multipath, location estimates do not meet any FCC requirement, in contrary to the hybrid data fusion. The developed hybrid TOA-AMC data fusion methodology even with SISO antenna mode the FCC requirements for network-based location systems are satisfied. MIMO 2x2 and 4x4 antenna modes introduce higher accuracy to the average RMSE. Only TOA MIMO 4x4 – AMC hybrid algorithm can be used for handset-base positioning systems as it meets FCC requirements.

Similar to the hybrid TOA-AMC data fusion methodology, TOA-DOA algorithm shows higher accuracy in comparison with TOA. However TOA-DOA average RMS is higher than for the TOA-AMC. TOA-DOA with SISO antenna mode (Figure 5.10) does not meet any FCC standards. As it can be observed in Figure 5.11 and 5.12, hybrid scheme meet the FCC requirements in the MIMO 2x2, and handset-based – in MIMO 4x4 antenna mode.

By analysing the results for the WIMAX network, the conclusion is that the MIMO 4x4 allows obtaining the most accurate position estimates. Hence, it is decided to present simulation results only for this case for the Heterogeneous

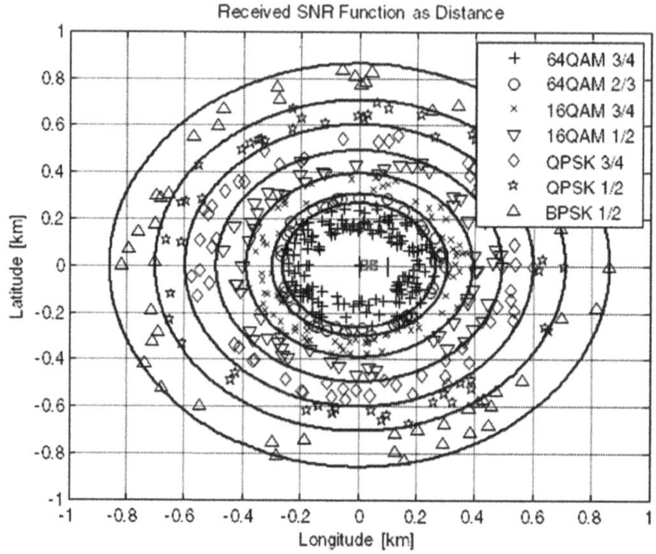

Figure 5.4 Received SNR Function as Distance in Cost 231 Hata Loss Model.

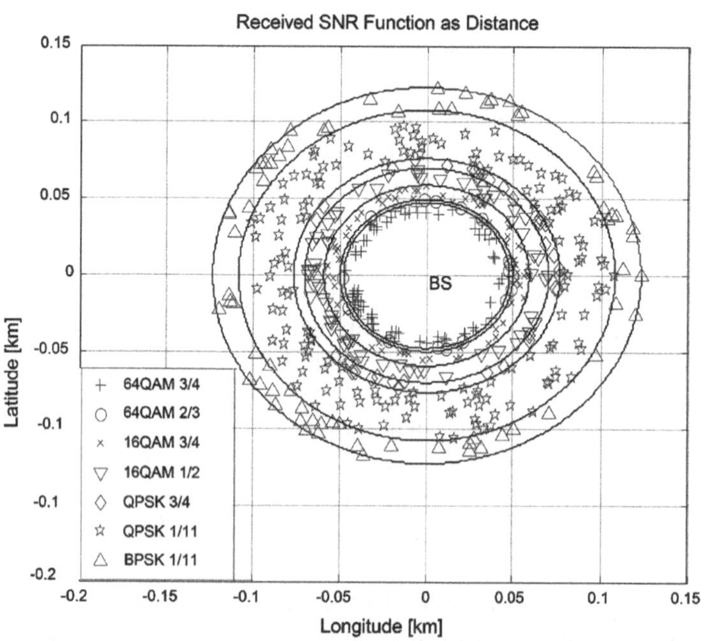

Figure 5.5 Received SNR Function as Distance in Cost 231 Hata Loss Model in WiFi.

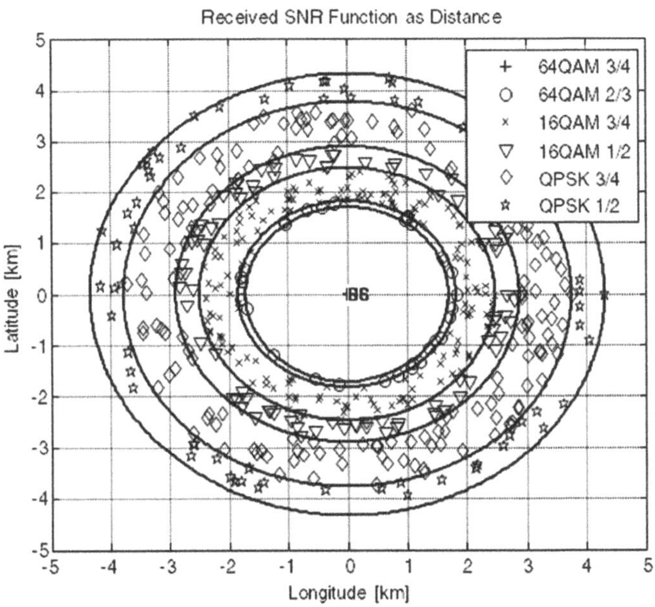

Figure 5.6 Received SNR Function as Distance in Cost231 Hata Loss Model in LTE.

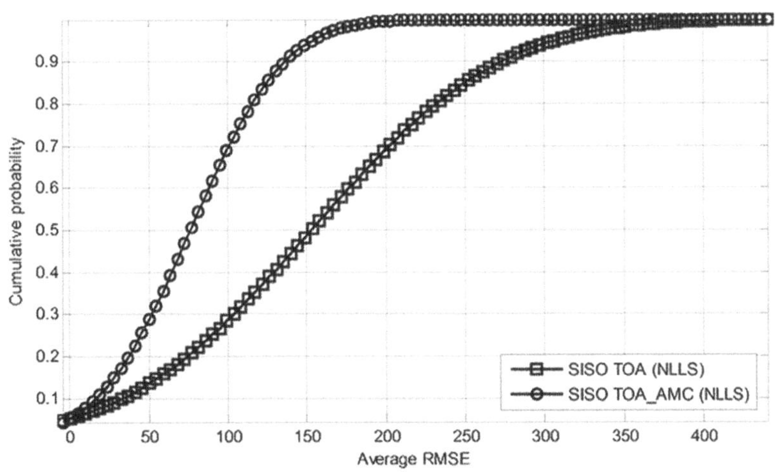

Figure 5.7 Comparison of the SISO TOA average RMSE with the hybrid TOA-AMC data fusion.

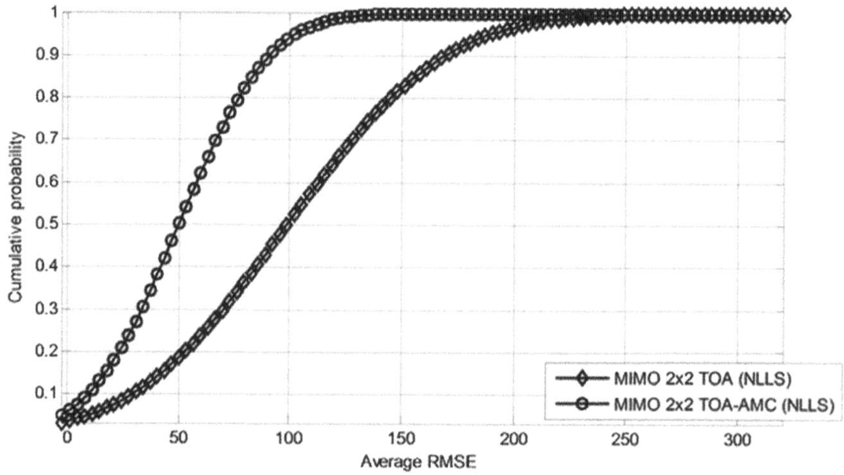

Figure 5.8 Comparison of the MIMO 2x2 TOA average RMSE with the hybrid TOA-AMC data fusion.

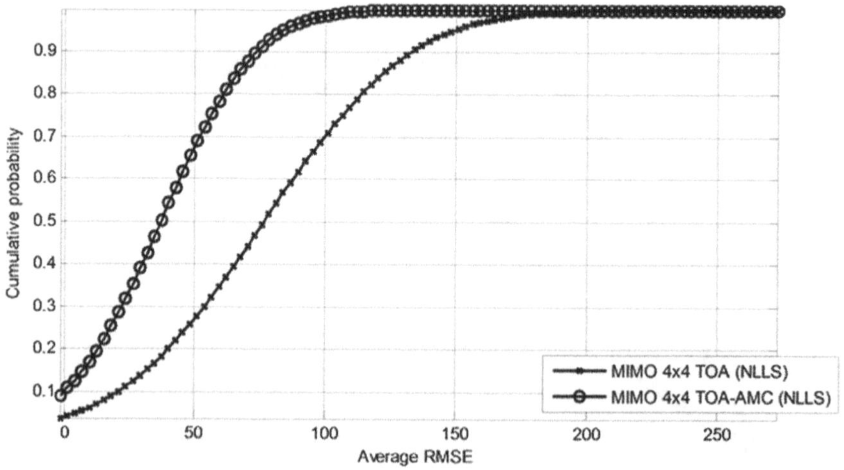

Figure 5.9 Comparison of the MIMO 4x4 TOA average RMSE with the hybrid TOA-AMC data fusion.

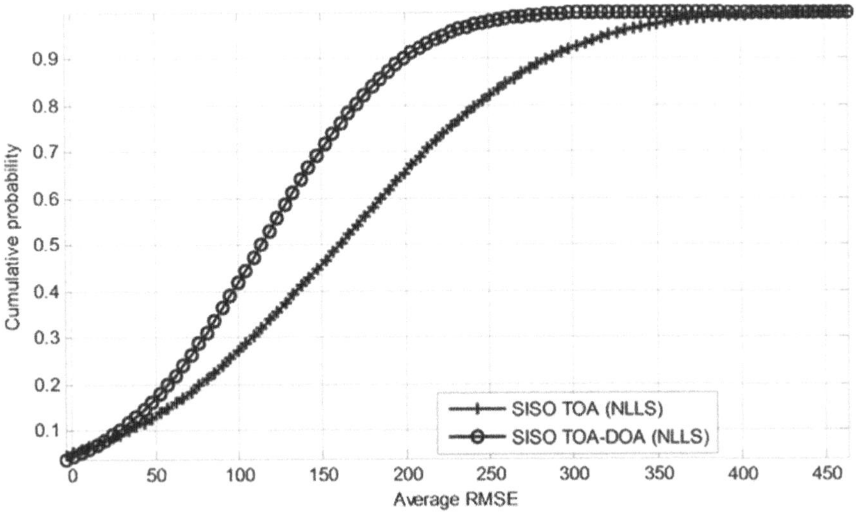

Figure 5.10 Comparison of the SISO TOA average RMSE with the hybrid TOA-DOA data fusion.

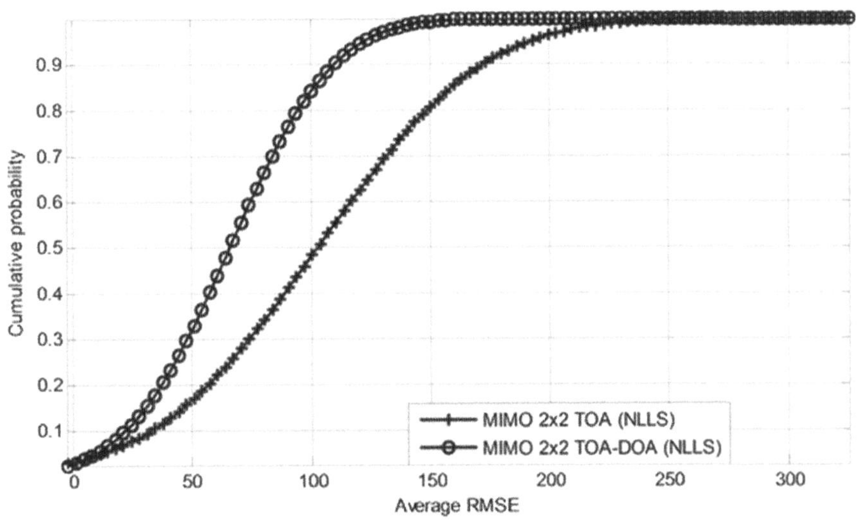

Figure 5.11 Comparison of the MIMO 2x2 TOA average RMSE with the hybrid TOA-DOA data fusion.

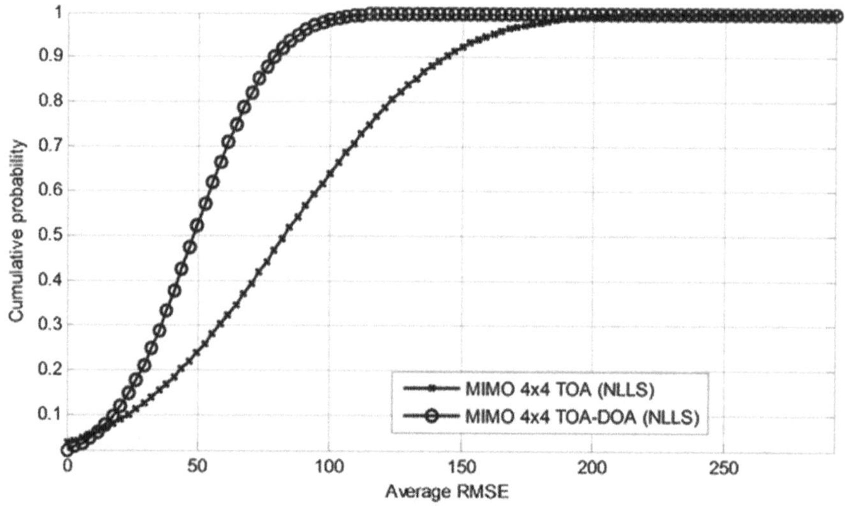

Figure 5.12 Comparison of the MIMO 4x4 TOA average RMSE with the hybrid TOA-DOA data fusion.

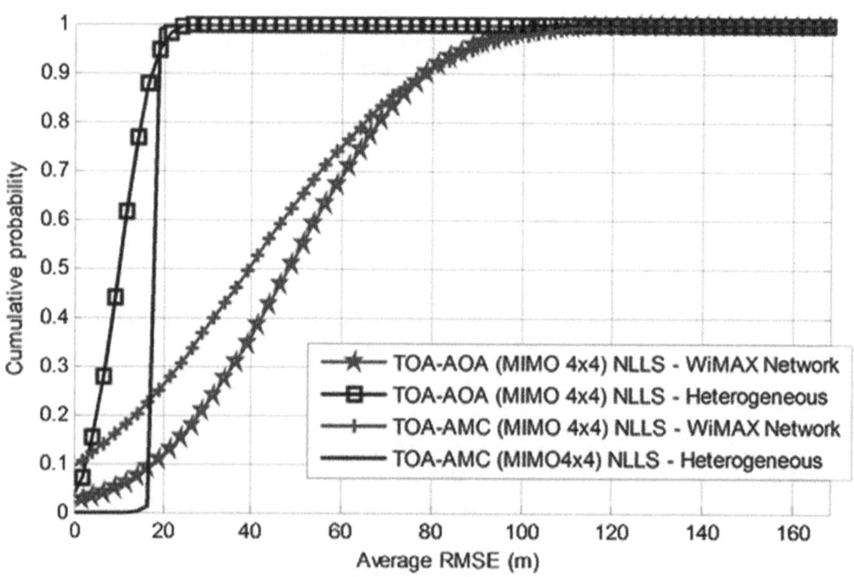

Figure 5.13 Comparison of the MIMO 4x4 Hybrid TOA-DOA and TOA-AMC schemes of the WiMAX and Heterogeneous Network Scenarios.

wireless network. In Figure 5.14 comparison of the presented positioning algorithms for the WIMAX and heterogeneous networks is presented. Hence, the comparative performance for the WiMAX and heterogeneous wireless networks are illustrate. As it can be seen the location error is minimized while using the heterogeneous wireless as WiFi has smaller area of coverage and signal is less corrupted.

Conclusion

This chapter presents methods and solutions that make possible to overcome drawbacks of GPS by utilizing 4G and 3G technologies for L&P. All possibly available wireless networks technologies and its signal measurements that can be used for positioning services are utilized. A proposed novel method of cooperation to perform positioning services in the mobile device based on coexistence of WiFi, WiMAX and LTE technologies is tested and analysed.

References

[1] S. Sand, C. Mensing, A. Dammann. Positioning in Wireless Communications Systems – Introduction and Overview, Proceedings 18th Wireless World Research Forum (WWRF) Meeting, Espoo, Finnland, June 2007.

[2] J. Zhou, J.Shi. RFID Localization Algorithms and Applications – a review. Journal Intelligent Manufacturing, Vol. 20, No. 6, pp. 659–707.

[3] L. Aalto, N. Göthlin, J. Korhonen, T. Ojala, Bluetooth and WAP Push Based Location-Aware Mobile Advertising System. Proceedings of the 2nd International Conference on Mobile Systems, Applications, and Services (MobiSys'2004), pp. 49–58, 2004.

[4] H. Oiso, M. Kishimoto, Y. Takada, T. Yamazaki, N. Komoda, T. Masanari, A Bluetooth-based Guidance System In-building Location Estimation Method. Application and Service in Wireless Networks, pp. 141–143, Milford, CT, USA, Kogan Page, Ltd., 2002.

[5] N.S. Correal, S. Kyperountas, Q. Shi, M. Welborn, An UWB Relative Location System, Proceedings of the IEEE Conf. on Ultra Wideband Systems and Technologies, pp. 394–397, November 2003.

[6] Y.-H. Jo, J.-Y. Lee, D.-H. Ha, S.-H. Kang, Accuracy Enhancement for UWB Indoor Positioning Using Ray Tracing, Proceedings of the IEEE/ION Position, Location and Navigation Symposium, pp. 565–568, 2006.

[7] S. Venkatesh, R.M. Buehnerer, Power Control in UWB Position-Location Networks, Proceedings of the IEEE International Conf. on Communications, Vol. 9, pp. 3953–3959, 2006.

[8] K. Yu, J.P. Montillet, A. Rabbachin, P. Cheong, I. Oppermann, UWB Location and Tracking for Wireless Embedded Networks, Signal Processing, Vol. 86, No. 9, pp. 2153–2171, 2006.

[9] G.Gomes, H.Sarmento. Indoor Positioning System Using ZigBee Technology. Proceedings of the Third International Conference on Sensor Technologies and Applications, pp.152–157, 2009.

[10] H. Cho, M. Kang, J. Kim, H. Kim, ZigBee Based Location Estimation in Home Networking Environments. IEICE Transactions on Information and Systems, Vol. E90-D, No. 10, pp. 1706–1708, 2009.

[11] Q. Yao, F.-Y. Wang, H. Gao, K. Wang, H. Zhao, Location Estimation in ZigBee Network Based on Fingerprinting, Proceedings of the IEEE International Conference on Vehicular Electronics and Safety, ICVES, December 2007.

[12] K. Kaemarungsi, P. Krishnamurthy, Modelling of Indoor Positioning Systems Based on Location Fingerprinting, Proceedings of the 23th Annual Joint Conf. of the IEEE Computer and Communications Societies, INFOCOM, Vol. 2, pp. 1012–1022, 2004.

[13] Z. Xiang, S. Song, J. Chen, H. Wang, J. Huang, X. Gao, A Wireless LAN-based Indoor Positioning Technology, IBM Journal of Research and Development, Vol. 48, N. 5/6, pp. 617–626, 2004.

[14] C. L. F. Mayorga, F. Della Rosa, S. A. Wardana, G. Simone, M. C. N. Raynal, J. Figueiras, S. Frattasi, Cooperative Positioning Techniques for Mobile Localization in 4G Cellular Networks, Proceedings from the IEEE International Conference on Pervasive Services, pp. 39–44, 15–20 July 2007.

[15] P. Castro, P. Chiu, T. Kremenek, R. Muntz. A Probabilistic Room Location Service for Wireless Networked Environments, In Proceedings of Ubicomp'01, October 2001, pp. 18–27. 2001.

[16] A. Smaliagic, D. Kogan, Location Sensing and Privacy in a Context-aware Computing Environment. IEEE Wireless Communications, Vol. 9, No. 5, pp. 10–17. 2002.

[17] P. Bahl, V. N. Padmanabhan, RADAR: An In- Building RF-Based User Location and Tracking System, Proceedings of 19th Annual Joint Conf. of the IEEE Computer and Communications Societies, INFOCOM 2000, March 2000, pp. 775–784.

[18] P. Myllymaki, et.al. A Probabilistic Approach to WLAN User Location Estimation. Proceedings of the third IEEE Workshop on Wireless LANs, September 2001, pp. 59 – 69.

[19] R. Battiti, T. L. Nhat, and A. Villani. Location-Aware Computing: A Neural Network Model for Determining Location in Wireless LANs. Technical Report DIT-02- 0083, University of Trento, Italy, 2002.

[20] A. M. Ladd, K. E. Bekris, G. Marceau, A. Rudys, D. S. Wallach, E. E. Kavraki, Using Wireless Ethernet for Localization, Proceedings of the IEEE/RSJ International Conference on Intelligent Robots and Systems (IROS)), September 2002, Vol. 1, pp. 402–408.

[21] A. Howard, S. Siddiqi, and G. S. Sukhatme, An Experimental Study of Localization Using Wireless Ethernet, Proceedings of the 4th International Conference on Field and Service Robotics, July 2003, pp. 145–153. 2003.

[22] A. Smailagic, D. P. Siewiorek, J. Anhalt, D. Kogan, and Y. Wang, Location Sensing and Privacy in a Context Aware Computing Environment, Proceedings of the International Conference on Pervasive Computing, May 2001, pp. 15–23.

[23] P. Myllymaki, T. Roos, H. Tirri, P. Misikangas and J. Sievänen Graphical Model on Manhattan: A Probabilistic Approach to WLAN User Location Estimation, International Journal of Wireless Information Networks, Vol. 9, No. 3, 2002, pp. 155–164.

[24] Ekahau, Inc., 12930 Saratoga Avenue, Suite B-9, Saratoga, CA 95070, 1–866-435–2428; see www. ekahau. com.

[25] Y.-B. Lin, Y.-C. Lin, WiMAX Location Update for Vehicle Applications, Mobile Network Applications, Vol. 15, No. 1, pp.148–159. 2010.

[26] S. S. R. Ahmed, Technological Strategy of Using Global Positioning System: an Analysis, International Journal of Engineering Science and Technology, Vol. 1, No. 1, pp.6–16, 2008.

[27] F. Della Rosa, Cooperative Mobile Positioning and Tracking in Hybrid WiMAX/WLAN Networks, MSc Thesis, Aalborg University. 2007.

[28] P. Vorst, J. Sommer, C. Hoene, P. Schneider, C. Weiss, T. Schairer, W. Rosenstiel, A. Zell, G. Carle, Indoor Positioning via Three Different RF Technologies, Proceedings of the 4th European Workshop on RFID Systems and Technologies. June 2008, Freiburg, Germany.

[29] N. Patwari, J. N. Ash, S. Kyperountas, A. O. Hero, R. L. Moses, N. S. Correal, Locating the Nodes: Cooperative Localization in Wireless

Sensor Networks, IEEE Signal Processing Magazine, Vol. 22, No. 4, pp. 54–69. July 2005.

[30] M. K. Widyawan, D. Pesch, A Bayesian Approach for RF-Based Indoor Localisation, Proc. of the International Symposium on Wireless Communication Systems, 2007, pp. 133–137.

[31] C.-S. Chen, Y.-J. Chiu and J.-M. Lin, Hybrid TOA/AOA Schemes for Mobile Location in Cellular Communication Systems. International Journal of Ad hoc, Sensor & Ubiquitous Computing (IJASUC). Vol. 1, no. 2, pp. 54–65, 2010.

[32] A. A. Md Isa, Enhancing Location Estimation Accuracy in WiMAX Networks. Proceedings of the 15th IEEE Mediterranean Electrotechnical Conference, pp. 725–731, 2010.

[33] P. Bahl, V. N. Padmanabhan, RADAR: An In- Building RF-Based User Location and Tracking System, Proceedings of the 19th Annual Joint Conf. of the IEEE Computer and Communications Societies, INFOCOM 2000, Vol. 2, pp. 775–784, March 2000.

[34] Intel Support, "Wireless Networking–What is Multiple-Input/ Multiple-Output (MIMO)?". Available: http://intel. com/support/wireless/sb/cs-025345.htm.

[35] J. G. Andrews, A. Ghosh, R. Muhamed, Fundamental of WiMAX: Understanding Broadband Wireless Networking. Prentice Hall, 2007.

[36] S. R. Saunders, A. A. Zavala, Antennas and Propagation for Wireless Comunication Systems, Wiley, 2005.

[37] S. Ranvier, Path Loss Models. Helsinki University of Technology, SMARAD Centre of Excellence. Online presentation. 2004. Available: http://www. comlab. hut. fi/opetus/333/2004_2005_slides/ Path_loss_models. pdf.

6

The Software Communications Architecture

Todor Cooklev
Wireless Technology Center
Indiana University – Purdue University Fort Wayne,
Fort Wayne, Ind., USA
E-mail: Todor Cooklev <cooklevt@ipfw.edu>

Anton Hristozov
Network Appliance
Pittsburgh, Pennsylvania, USA
E-mail: antonhr@verizon.net

6.1 Introduction

The United States military supports a very wide range of communication devices. These devices range from simple man pack radios to sophisticated wireless systems on vehicles, aircraft carriers and nuclear submarines. For example, SINCGARS, designed and built by ITT Exelis, Inc. of Fort Wayne, Indiana, is shown in Figure 6.1. Due to its operational effectiveness SINCGARS has become the most widely deployed tactical radio in the world today. Figure 6.2 shows another system, a Joint Tactical Radio System (JTRS)-compliant Rifleman radio.

These systems are different, but especially during combat operations they must communicate among themselves. Furthermore these systems are desirable to interoperate with the communication systems of coalition forces and allies. This requirement creates enormous challenges. This is a main reason why the US DoD has been an important proponent for software-defined radio (SDR) technology in general. In particular, the DoD has decided that the backbone of military communications is the Joint Tactical Radio System, and its architecture is the Software Communications Architecture (SCA). Today, the intention of the US DoD is to procure only SCA-compliant radios. JTRS is a procurement program unique in its interoperability requirement to ensure

Vladimir Poulkov and Ramjee Prasad (Eds.), Resource Management in Future Internet, 143–184.

Figure 6.1 SINCGARS radio (printed with permission of ITT Exelis, Fort Wayne, Ind).

Figure 6.2 Rifleman radio (printed with permission of ITT Exelis, Fort Wayne, Ind).

that any JTRS-compliant hardware can support any JTRS-compliant software. The SCA is also beginning to be supported by military organizations in other countries around the world.

To understand the need for architecture like the SCA, it is important to note that software-defined radios are usually implemented using several types of digital hardware devices, such as general-purpose processors (GPPs), digital signal processors (DSPs), and field-programmable gate arrays (FPGAs). These different hardware technologies are required to satisfy the timing and functionality demands of a high-performance solution. DSP's are typically used to implement low speed signal processing portions of the baseband subsystem. FPGAs implement higher-speed signal processing portions. Applications are implemented on GPPs. Developing solutions that run on several hardware types is not straightforward. The SCA accomplishes these tasks by defining a set of interfaces that isolate the applications from the hardware. This set of interfaces is referred to as the Core Framework of the SCA. Additionally, the SCA ensures that once software components are deployed on a system, they are able to execute and communicate with the other hardware and software elements present in the system. As a result, an SCA application consists of multiple software components that are loaded onto the appropriate processing resource.

To understand the SCA we must differentiate between waveform software – software that implements signal processing operations – from the software that provides the capabilities for the waveform software to execute. The former can be called "core" radio software or just functional software, and the latter – management software. The SCA is concerned with the management software. The role of the SCA is to provide a common infrastructure for managing the software and hardware elements and ensuring that their requirements and capabilities are commensurate.

From a software development perspective, the SCA is a Component-Based Development (CBD) architecture, i.e. it is an architecture for the creation and integration of software components. An application is assembled using software components much like a system is built using hardware components. CBD is the most popular programming paradigm, used for example within Microsoft's .NET.

A general illustration of the SCA is shown in Figure 6.3. The architecture has been developed using an object-oriented approach. The software components which provide for the management and execution of the SCA comprise the operating environment (OE). The OE consists of an operating system (OS), and the elements defined by the Framework Control and Service Interfaces. Communications between the application and the Framework Services Interfaces are accomplished through the Transfer Mechanism (which in previous versions of the SCA used to be the Common Object Request Broker Architecture (CORBA), including the Event and Naming Services).

The SCA has three main segments: Waveform, Core Framework, and the Domain Profile. The CF has all components to manage the radio system and deploy applications. The Domain Profile includes the XML files that describe the hardware resources within the system.

There are several incorrect perceptions about the SCA [1]. Probably the most significant is that the SCA Operating Environment (CORBA middleware and POSIX) requires substantial resources and adds too much overhead. This perception is based on TCP/IP being used as the transfer mechanism by the CORBA middleware. The SCA Next was developed in response to this concern. Some other myths about the SCA are that it is an implementation

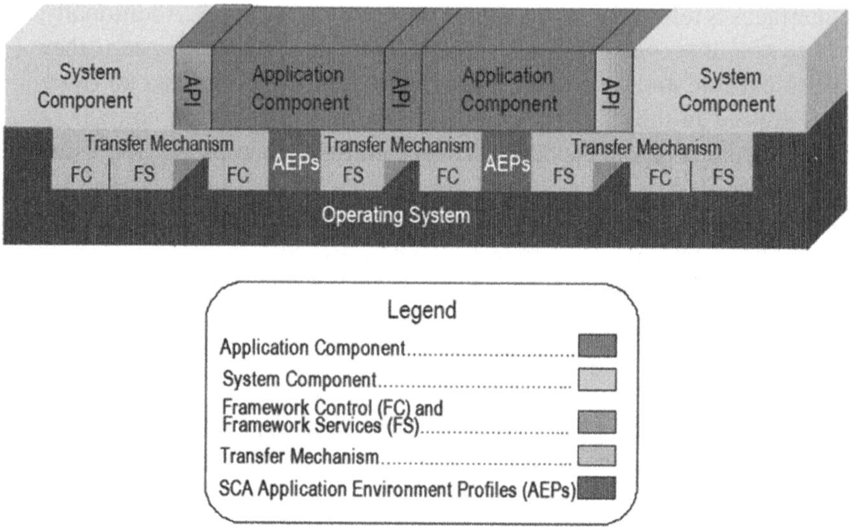

Figure 6.3 SCA Architecture.

architecture; that it applies only to the frequency range below 2 GHz, etc. In reality it is implementation-independent and there is nothing that limits the frequency range of SCA radios. The SCA is technology-independent. It imposes constraints only on the interfaces and on the structure of the software. IDL is used to define the SCA interfaces. Unified Modelling Language is used to graphically represent SCA interfaces, operational scenarios, and use cases. It supports plug-and-play so that hardware and software devices from different vendors can work together.

In this Chapter, Section 6.2 is devoted to the diverse set of technologies that are relevant for the SCA. Section 6.3 is devoted to the Operating Environment and the Core Framework, Section 6.4 – to the latest version called SCA Next. In Section 6.5 we discuss specifics for DSP, FPGA, and GPPs and Section 6.6 contains design principles that should be followed in the development process. The Chapter concludes with Section 6.7.

6.2 Hardware and Software Technologies Used in the SCA

Today's SDR devices are complex distributed systems which require parallel processing and appropriate real time communication.

Some modern wireless systems can be considered to be networks of computers and some can be considered distributed systems. In a distributed system, a collection of independent computers appears to its users as a single coherent system. Conversely, in a computer network, this coherence is absent. Users are exposed to the actual machines, without any attempt by the system to make the machines look and act in a coherent way. If the machines have different hardware and different operating systems, that is fully visible to the users. If a user wants to run a program on a remote machine, he has to log onto that machine and run it there. In effect, a distributed system is a software system built on top of a network. The software gives it a high degree of cohesiveness and transparency. Thus, the distinction between a network and a distributed system lies with the software, rather than with the hardware. A well-known example of a distributed system is the World Wide Web.

SDR's can be conceived of as a network of computers. In some implementations, they actually are; with separate processors for each subsystem. Developing applications running across these devices in a heterogeneous environment is difficult. Note that heterogeneity occurs at different levels:

- **Programming languages:** Different objects can be developed in different programming languages.
- **Operating systems:** Different operating systems may be used that have different characteristics and capabilities.
- **Computer architectures:** Computers differ in their technical details (e.g., data representations and endianness, as well as CPU type – DSP, FPGA or GPP).
- **Networks:** Different computers are linked together through different network technologies.

It must be noted that there are no universally appropriate hardware and software design approaches. This is difficult because of the multitude of diverse hardware components used to build radios and the accompanying many diverse software environments. While there are no general design approaches, at the system architecture level there several potential choices from which the designer may choose. These potential choices are examined in this section.

6.2.1 OOD

Object-oriented design is the method which leads to software architectures based on the objects every subsystem manipulates. Clearly, the main design question is not what a given system "does", but what data it does it to. Using the object-oriented design, one is almost forced to postpone answering the question about the topmost function. For communication engineers this change in approach may be as much of a shock. This, however, is the key to software re-usability.

Object-oriented design is an evolution of structured design. When object-oriented design is used for software systems, it is called object-oriented programming. The program is based on modifying the variables of these objects and program execution is based on the interaction among objects. This approach is better suited to event-driven processing than the linear model. Programming languages like C++ were designed to implement the Object-Oriented Programming (OOP) model, however OOP can be implemented with any language. In fact, there is a school of thought that says that developers should learn C++ in order to master OOP and then implement the Object-Oriented Programming model in C, thereby avoiding the overhead (and run time speed reduction) of C++. Because of the speed of execution issues in SDR, this school of thought is particularly prominent in SDR developers.

In the Object-Oriented Programming Model, every object contains two types of elements: variables and methods. A variable is simply an entity that contains one value or a set of values. A method is a function that can modify variables, access other methods or objects. Depending on how they are declared, variables and methods may be used by:

- other methods that are part of the same object,
- other methods that belong to a select group of objects,
- or any other method.

The functionality of an object is isolated from outside, a.k.a. encapsulation. Therefore the functionality of an object can evolve, as long as the interface with the object remains the same. This principle is also known as design by contract. Objects are declared just like variables, but their types are called *classes*. An *object* represents an instance of a class. In general, objects can be created and destroyed.

There are two main advantages to object-oriented design: modularity and data hiding. Modularity makes it possible for the entire system to be broken into separate, independent pieces. In this way maintenance, re-use, and portability are simplified. Objects can change independently. The controlled access to variables leads to data hiding. Data encapsulation is an advantage because it hides the implementation from the interface. This gives an opportunity to change the internal implementation as the software evolves but to keep the same interface and thus keep the software components loosely coupled which is a desirable feature in software architecture.

The communication among objects is very important. This communication is simple for objects operating in the same environment and is an inherent part of the programming language. Other important concepts are inheritance, and polymorphism. Inheritance represents the ability to reuse types. This way the subclasses can inherit from the super class and add features specific to them just like organisms do in evolution. Polymorphism is the reimplementation of the same operation for different types, which is allowed by the type hierarchy. Polymorphic behavior means that an operation with the same name can do something totally different depending on the current object context.

6.2.2 UML

One of the main approaches used in designing object oriented software today is through UML. The OMG's Unified Modelling Language^{TM} (**UML®**) helps the user specify, visualize, and document models of software systems, including their structure and design, in a way that meets all of the requirements. One can model just about any type of application, running on any type and combination of hardware, operating system, programming language, and network, in UML. Its flexibility lets you model distributed applications that use just about any middleware on the market. Built upon fundamental object oriented concepts including *class* and *operation*, it's a natural fit for object-oriented languages and environments such as C++ or Java.

UML 2.0 defines thirteen types of diagrams, divided into three categories: Six diagram types represent static application structure; three represent general types of behavior; and four represent different aspects of interactions:

- **Structure Diagrams** include the Class Diagram, Object Diagram, Component Diagram, Composite Structure Diagram, Package Diagram, and Deployment Diagram.
- **Behavior Diagrams** include the Use Case Diagram (used by some methodologies during requirements gathering); Activity Diagram, and State Machine Diagram.
- **Interaction Diagrams,** all derived from the more general Behavior Diagram, include the Sequence Diagram, Communication Diagram, Timing Diagram, and Interaction Overview Diagram.

The most popular diagrams in UML are the class diagrams, the sequence diagrams, state diagrams and the activity diagrams. For real-time systems the timing diagram is also used.

The class diagram is a static diagram which represents the relationships between classes in the class hierarchy of the object oriented model. An example of two classes inheriting from one base class is given in Figure 6.4.

The arrow signifies generalization which is the functionality encapsulated in the Basic Radio class. All specific examples inheriting from the base class are adding specific functionality to the already defined functionality by the base class. This achieves code reuse and interface reuse.

Activity diagrams are used to represent algorithms. All the steps or activities are captured in the state boxes. They are useful to represent a sequence of steps in a language independent form. An activity diagram example of a start up sequence is given in Figure 6.5.

Activity diagrams allow parallel activities to be represented which is very useful in multi-processor and multi-tasking environments. They are excellent choice when dynamic behavior needs to be captured in a concise and illustrative way.

Two more complete examples of class diagrams follow.

UML can be used in the design of hardware using an object-oriented approach normally used for software design. Following this method the hardware can be

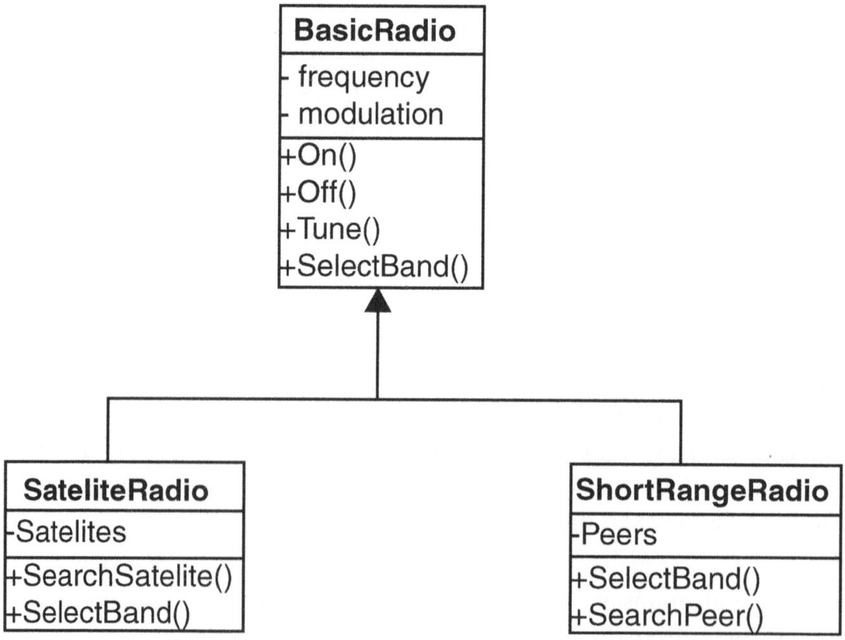

Figure 6.4 Radio Hierarchy.

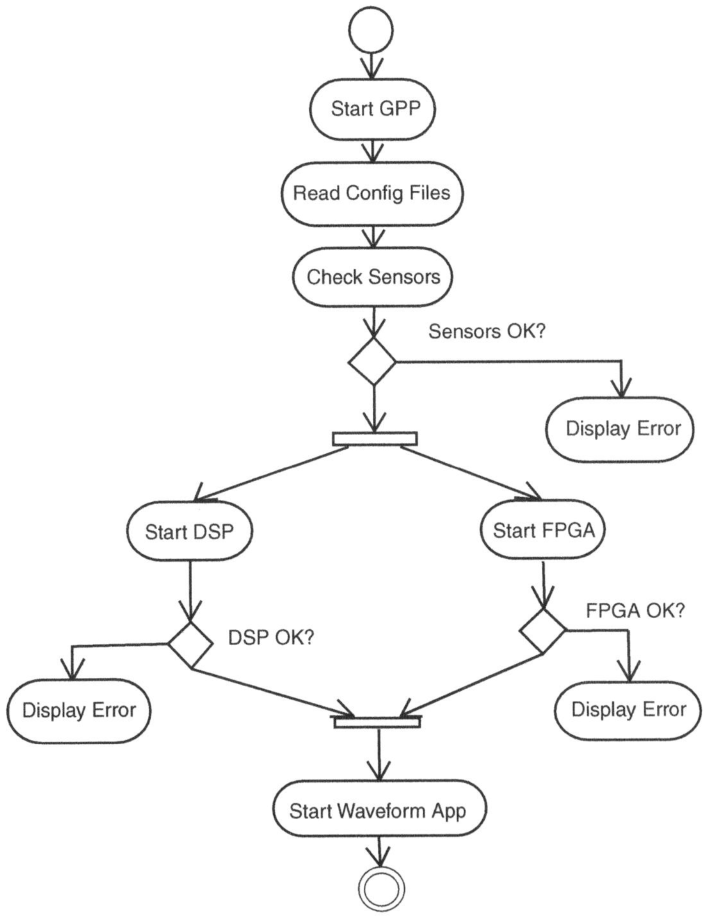

Figure 6.5 Startup Sequence of a Software Defined Radio.

viewed as a hierarchy of hardware-class and subclass objects. The class structure shows how object-oriented classes and subclasses are related. The overall hardware parent is the *SCA-Compliant Hardware* class; it has attributes such as maintainability, availability, physical, environmental, and device registration parameters. *SCA-Compliant Hardware* has two child classes: *Chassis* and *HW Modules*. Each of the hardware child classes can be extended further. The OO concept of inheritance allows devices to inherit from their parents and share common physical and interface attributes. This is known as "is-a" relationship in object oriented design. For example each of Antenna, Receiver and Power-Amplifier is a RF descendent as each class adds something specific to the ancestor RF. For example, as shown in Figure 6.6 the *RF* class is extended by the addition of *Antenna*, *Receiver*, and *Power Amplifier* child classes. The modem class has modulator and demodulator subclasses and can support multiple air interfaces as shown in Figure 6.7. In addition to inheritance classes can contain other classes and then we have composition which represents has-a relationship. UML is now a de facto standard used for software design of different types of software including object and non-object oriented systems. Because of its universality UML is also applicable for design of SDR systems. The knowledge of UML in the industry is spreading throughout the practicing engineers and thus UML is becoming very well understood and this further eases its wide adoption.

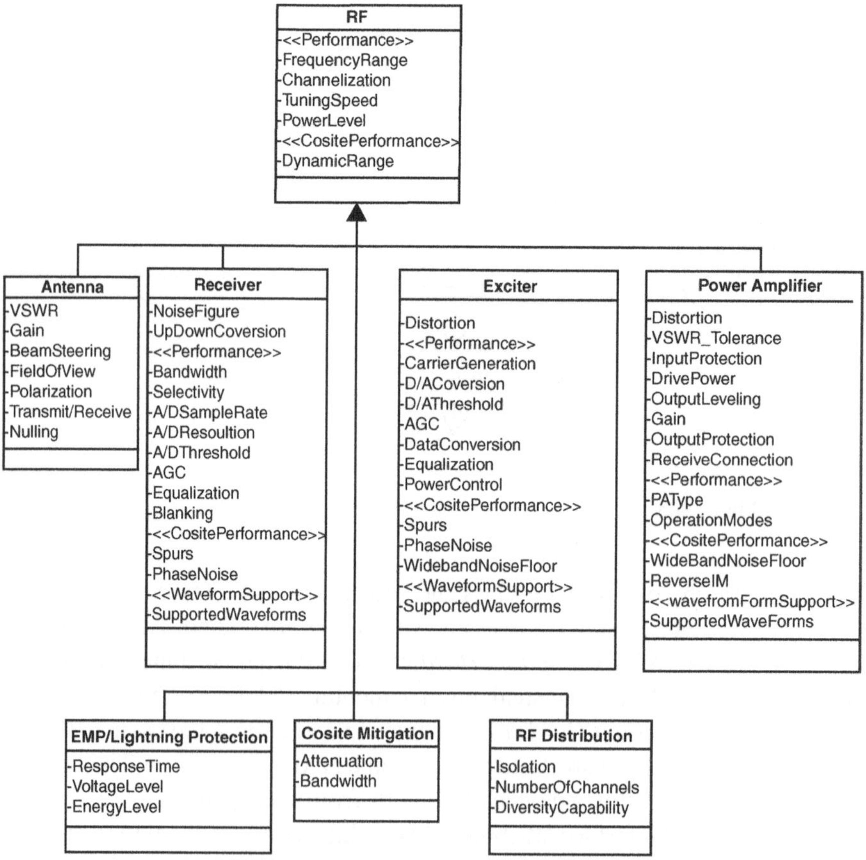

Figure 6.6 UML class diagram for an example radio class hierarchy.

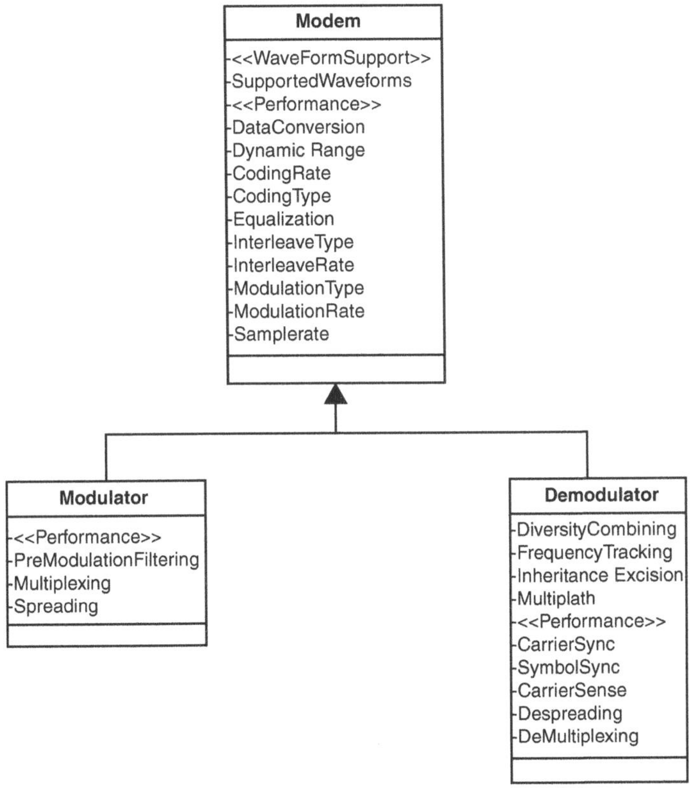

Figure 6.7 UML class diagram for a modem class hierarchy.

6.2.3 XML

Extensible Markup Language (XML) is an application of the Standard Generalized Markup Language (SGML)[1]. XML is an explicit structure for the description of information. It is easy for humans to interpret and easy for machines to process. It provides explicit structure for the description of information. It has two parts – prolog and properties. The prolog contains structural information. This structural information is also called DTD (document type definition) and describes what the document should look like. Elements are marked with tags such as "<element>value" and can be nested. The XML parser checks the syntax and grammar of XML files. In the SCA, XML files

[1]Note that like its name suggests HTML also has its roots in SGML

describe the layout of the system and the waveform applications, their location, names, etc.

6.2.4 Middleware

Since the different components used to make a SDR product are very different we need to bridge the software running on each of them. A layer of software, called middleware, overcomes this heterogeneity by offering identical functionality at all interface points. Middleware was developed to define interfaces among software modules that were not designed to work together. In general, middleware provides a certain level of translation between different pieces of software and different pieces of hardware. It provides the compatibility in parameters such as operating system, hardware platforms and protocol layers.

The term middleware suggests that it is software positioned between the operating system and the application. Viewed abstractly, middleware can be envisaged as a "tablecloth" that spreads itself over a heterogeneous network concealing the complexity of the underlying technology from the application being run on it (see Figure 6.8).

Middleware can also be considered to be between the applications and the underlying network. This layer provides various services like identification, authentication, naming, trading, security and directories. Middleware also aims to provide hardware and location transparency to software entities.

Middleware enables the interaction and communication between different applications through Application Programming Interfaces (APIs) across the distributed components.

The goal of middleware is to simplify the construction of a distributed system, where the engineers can abstract themselves from the implementation of low-level details. Middleware usually provides the following benefits to the software architecture:

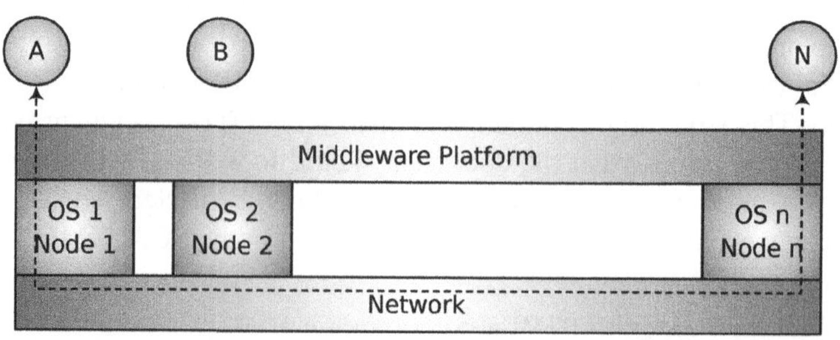

Figure 6.8 Illustration of Middleware Functionality.

- Hides distribution, i.e. the fact that an application is usually made up of many interconnected parts running in distributed locations.
- Hides the heterogeneity of the various hardware components, operating systems and communication protocols.
- Provides uniform, standard, high-level interfaces to the application developers and integrators, so that applications can be easily composed, reused, ported, and made to interoperate.
- Supplies a set of common services to perform various general purpose functions, in order to avoid duplicating efforts and to facilitate collaboration between applications.

There are three main types of middleware: object-based, event-based, and message-oriented. Message-oriented middleware works on the concepts of asynchronous message passing. It stores the messages sent to the client till they are acted upon. This type of middleware might perform some message transformation, i.e. the sent and received messages might not be the same. Object-based middleware works in an object-oriented environment and is the most popular. It can be implemented using object brokers.

The Common Object Request Broker Architecture (CORBA) is an example of an object oriented middleware. CORBA is an open distributed object computing infrastructure being standardized by the Object Management Group. It allows a distributed, heterogeneous collection of objects to interoperate. CORBA defines the architecture for distributed objects. The basic CORBA paradigm is that of a request for services of a distributed object. Everything else defined by the OMG is in terms of this basic paradigm. The services that an object provides are given by its interface. Interfaces are defined in OMG's Interface Definition Language (IDL). Distributed objects are identified by object references, which are typed by IDL interfaces. A client holds an object reference to a distributed object. The object reference is typed by an interface. The Object Request Broker, or ORB, delivers the request to the object and returns any results to the client.

The ORB is the distributed service that implements the request to the remote object. It locates the remote object on the network, communicates the request to the object, waits for the results and when available communicates those results back to the client. The ORB implements location transparency. Exactly the same request mechanism is used by the client and the CORBA object regardless of where the object is located. It might be in the same process with the client, down the hall or across the planet. The client cannot tell the difference. The ORB implements programming language independence for the request. The client issuing the request can be written in a different programming language from the implementation of the CORBA object. The ORB does the necessary translation between programming languages. Language bindings are defined for all popular programming languages.

Software Objects can be created by multiple vendors. As with conventional object oriented models, an object is an identifiable, encapsulated entity that provides one or more services that can be requested by a client. To communicate these objects need an object broker. This object broker allows the creation of software that has common interface to all entities with which it may have to work. The ORB (Object Request Broker) is an independent piece of software that performs data management between two separate objects, as illustrated in Figure 6.9. Note that these objects can be on different platforms as well. For example, a software-defined radio is likely to integrate multiple platforms – DSP, Reconfigurable Logic, and general-purpose processors. Middleware is needed to support the transfer of information between these different platforms.When the remote application gets connected to the main server, the methods of the server become available for the application.

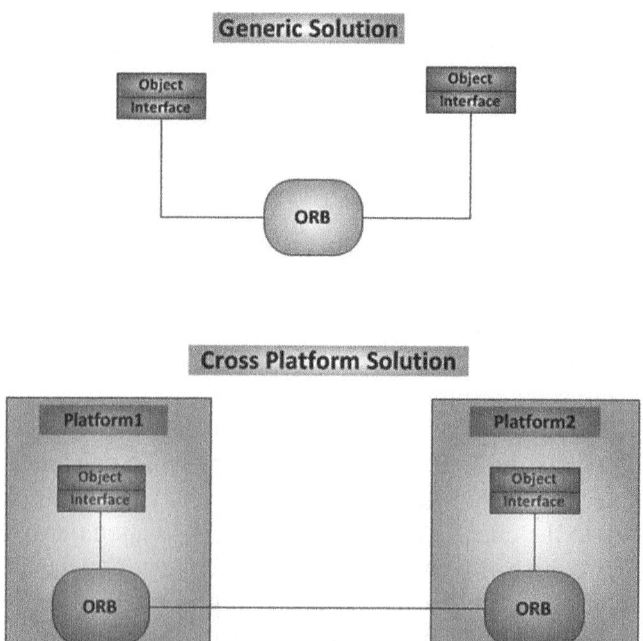

Figure 6.9 The functionality of object request brokers (ORB).

In the context of middleware a standard has to establish the interfaces between different components to enable their interaction with one another. We want to distinguish between two types of interface: a horizontal interface and a vertical one (Figure 6.10). The horizontal interface exists between an application and the middleware and defines how an application can access the functionality of the middleware. This is also referred to as an *Application Programming Interface* (API). The standardization of the interface between middleware and the application makes the the *portability* of the application easier. Thus the application can be ported to different middleware because the same API exists at each access point. The API depends on the programming language in which the application is written. This means that the API has to be adapted to the conventions of each programming language that is supported by the middleware. An application programmer typically perceives middleware as a program library and a set of tools. The details will naturally depend on the development environment that the programmer is using.

In addition to the horizontal interface, there is a vertical interface that defines the interface between two instances of a middleware platform. This vertical interface is typically defined through a protocol on the basis of messages, referred to as Protocol Data Units (PDU). A PDU is a message sent over the network. Both client and server exchange PDUs to implement the protocol. The vertical interface separates technological domains and

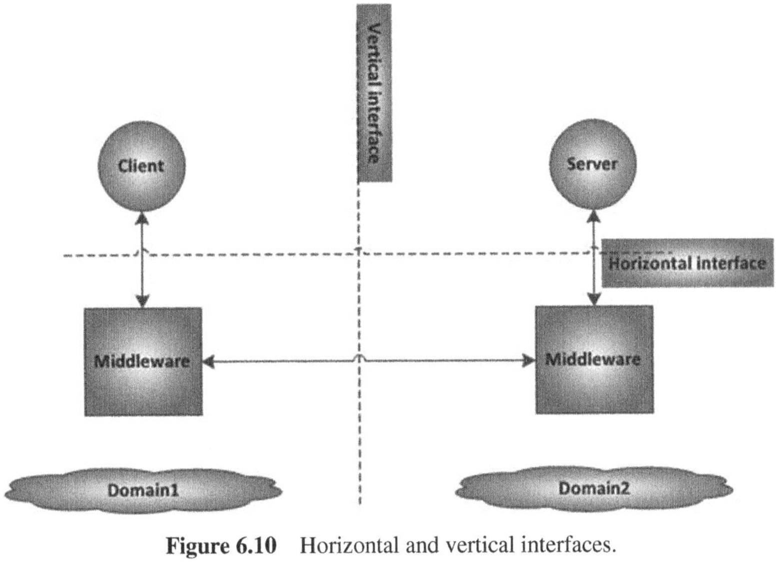

Figure 6.10 Horizontal and vertical interfaces.

ensures that applications can extend beyond the influence area of the product of middleware. The standardization of this interface allows *interoperability* between applications. This is illustrated in Figure 6.10.

CORBA (Common Object Request Broker Architecture) is an open system standard[2], which enables communication between distributed objects and supports the distributed execution of applications. The key component that allows this is the Object Request Broker. CORBA is one of the existing middleware standards which can operate across heterogeneous software entities. The latest SCA specification does not mandate CORBA for middleware as it was the case in previous SCA specifications. CORBA though seems to be one of the best options for this task.

An object in a network may request service from another object. This is accomplished by a service request by an object to the network infrastructure. The term "request" is broadly used to refer to the entire sequence of causally related events beginning with a client initiating a request and ending with the last event associated with the request. The client object is the object that requests a service by invoking an operation on an object implementation. The server object is the object that performs the service requested.

The Interface Definition Language (IDL), which is not a programming language, is used to write the interface of an object. An IDL compiler compiles the given IDL into *client stubs* and *object skeletons*. The stubs and skeletons act as proxies for clients and servers, and they run on top of Object Request Brokers. Each object in a CORBA application is defined as an interface, using IDL. The interface is the syntax part of the contract that the server object offers to the clients that invoke it. Any client that wants to invoke an operation on the object must use this IDL interface to specify the operation it wants to perform.

The client-side stub generates a request and sends it on behalf of a client object. The stub is the connection between the client and the communications link. The stub takes care of the marshalling of the actual parameters so that the request can be transmitted over the network properly. Marshaling is the process and the steps taken to represent an object in a form suitable for communications or storage. Similarly the reverse process un-marshaling happens when an object is received through a communication link and needs to be converted to its native form in memory or disk. The ORB is responsible for translating programming language function calls into network messages between clients and servers that provides transparent client/server

[2]CORBA is standardized by an industry association named OMG (Object Management Group)

relationships between objects. Note that as illustrated the ORB resides on each side.

Note that without an adaptor both the ORB and the objects must either agree on one fixed set of interfaces or support multiple sets of interfaces. Having a unique set of interfaces is not desirable because different sets of interfaces might be wanted, depending on the application's goals. Therefore the architecture contains an object adaptor. The ORB finds the object adapter that the server object is implemented in and passes the request on to the adapter. The object adapter finds the server.

When the invocation reaches the target object, the server-side skeleton (which is an object adapter) receives the request and delivers it to the CORBA object implementation. If the server uses a static skeleton the request is unpacked by the IDL-generated code and the desired method is invoked. The same interface definition is used there to un-marshal the arguments so that the object can perform the requested operation with them.

Results are sent by the server-side skeleton to the server-side ORB and then it is sent back to the client-side (see Figure 6.11). The interface definition is then used to marshal the results for their trip back, and to un-marshal them when they reach their destination.

Clients reach objects by using object references. The address of an object is encoded in a so-called Interoperable Object Reference (IOR) that contains the physical address as well as the name of the interface implemented by the object.

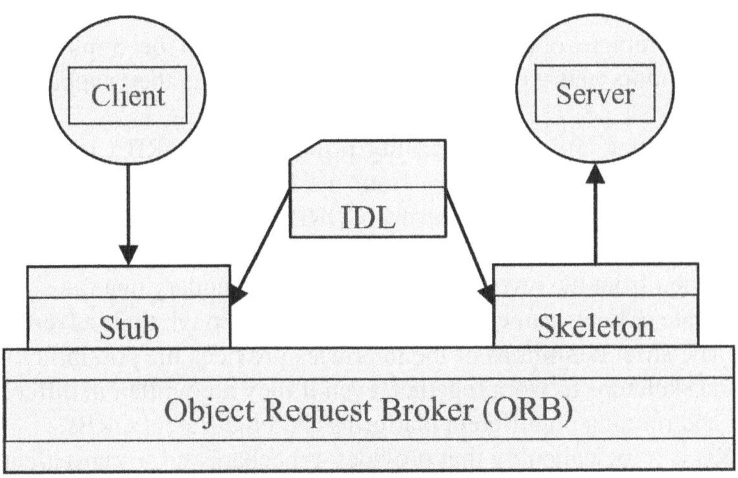

Figure 6.11 CORBA Stubs and Skeletons.

Every object has a unique object reference, and this reference can be obtained by a client in a number of ways. Once this reference is obtained, clients invoke operations on these objects, as if they are local objects. Actually, these operations are invoked on the client's stub, which then invokes the ORB on which the client is running. This ORB locates the ORB that has the real object implementation. The invocation continues through the target ORB, and the skeleton on the implementation side, to get to the object where it is executed. In this way a client object doesn't need to know where a server object is located, what platform it is running on, what communication protocol is needed to reach it, or what programming language it is implemented in. In CORBA all client invocations are routed by the ORB to a specific implementation called server classes. Based on the information contained within the object reference (or IOR), the local ORB will decide to invoke the operation either locally or remotely.

The horizontal interface introduced earlier in this section is defined in CORBA through IDL, whereas the vertical interface is a General Inter-ORB Protocol (GIOP). In a distributed network different ORBs communicate using this vertical interface. This protocol specifies a representation to specify the target object, operation, and all parameters (input and output) of every type that may be used. The GIOP specifies one main constraint: different ORBs must be TCP/IP compliant. This specialized form of the GIOP is referred to as the Internet inter-ORB Protocol (IIOP).

When the ORB that runs the client discovers that the actual object implementation is on a remote ORB, it routes the invocation out over the network to the remote object's ORB. The ORBs might be implemented by different vendors and will be interoperable as long as they support this interface.

The transparency and interoperability provided by CORBA is enabled by IDL. IDL separates the interface from the implementation. There is a very strict interface definition for every CORBA object. This interface is advertised throughout the system. In contrast, the object's running code and its data is hidden from the rest of the system with a boundary that the clients cannot penetrate. Clients can only reach the objects through their advertised interfaces. The strict definitions of the interfaces provides the possibility for the stubs and skeletons to work together, even if they are written in different languages, and running on different platforms and on different ORBs.

The ORB is a logical entity that provides a mechanism for transparently communicating client requests to target object implementations. The ORB simplifies distributed programming by decoupling the client from the detail of the method invocation. This makes client request appear to be local procedure calls.

An ORB may be implemented in various ways, such as a set of libraries. To decouple applications from implementation details, the CORBA specification defines abstract interfaces for an ORB. This interface provides various helper-functions such as converting object references to string and vice versa, and creating argument lists for requests made trough the dynamic invocation interfaces.

A CORBA system is a collection of objects that encapsulate data and functionality. In this way there is isolation between the request of services (clients) from the providers of services. The isolation is in terms of data representations and executable code and is provided by a well-defined encapsulating interface. From a client perspective, there is no special mechanism for creating or destroying an object, as it exists in an object oriented system. Objects are created and destroyed as an outcome of issuing requests.

Where CORBA services provide services for objects, CORBA facilities provide services for applications. These interfaces are also horizontally oriented, but unlike CORBA services they are oriented toward end-user applications. A client application program is written in a language supported by the ORB. Every ORB supports one single language mapping in which appropriate proxy are generated in this language.

Interfaces are the key concept of the CORBA object model. A client may request a set of possible operations from objects through their interfaces. Obviously, an interface may have several implementations. Application Interfaces are developed specifically for a given application. These interfaces are typically not standardized.

The ultimate goal of the CORBA architecture is to be neutral with respect to vendor, language platform, and language. This is achieved using the Interface Definition Language (IDL). The goal of CORBA is to enable interoperability between applications in heterogeneous distributed environments, without regards for where they are located. IDL is created for every object interface. Then an IDL "compiler" is used to generate source code in the respective language. The IDL-generated source code is compiled along with the rest of the project. The generated code is designed to interface with the ORB. A different IDL compiler is needed for every language to be supported. IDL is independent of compiler, software language, or operating system.APIs are described using IDL, standardizing the semantics of the interface. This guarantees that the semantics are compatible, through the API, and that the data is transferred correctly, through IDL.From this description, it becomes apparent that CORBA is considerably more complicated than a simple ORB. Figure 6.12 shows a block diagram of CORBA.

There is a naming service. It is a generic directory service that provides functionality analogous to the white pages, such as finding addresses of a person when his/her name is known.

A trading service allows service providers to advertise services and allows clients to search and locate available services. An object that provides trading services in a distributed system is called a “trader”. Clients are referred to as importers or consumers of information and servers are referred to as exporters or providers of information. In order to find service offer(s) which satisfy its required matching constraints, the service request is made with following details: service type, service properties, and servicing and scoping policies.

An *event service* allows components to dynamically register or unregister their interest in specific events. The service defines an object called an event channel that collects and distributes events among components that otherwise may know nothing about each other. Subscribers connected to the event channel receive the events asynchronously. This event channel provides a high degree of decoupling between clients and servers. There are several channels

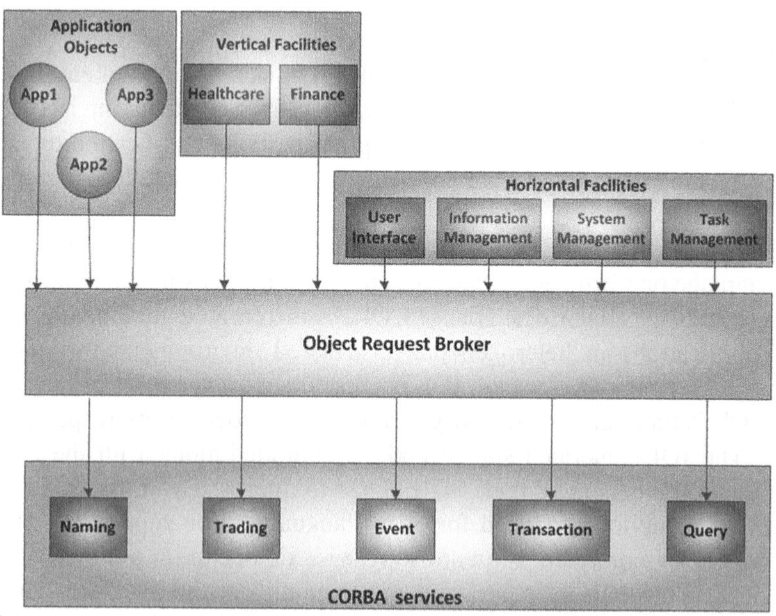

Figure 6.12 OMG Object Model Architecture.

of communication. All events received from suppliers are broadcast to all consumers over specific event channels, called untyped channels. Over the typed event channels communication is restricted only to the suppliers and consumers that are interested in a particular IDL interface.

The *object transaction service* controls the scope and duration of transactions. It allows multiple objects to participate in a transaction.

It becomes apparent that dealing with database aspects is extremely important for CORBA. A number of vendors deal with data sources and facilities to manipulate theses data sources. Database systems may not support a wide range of operating systems. Furthermore, some database systems do not support multiple languages. And if they do, then they do not allow interoperability between them.

CORBA, with the object request broker ORB, can provide communication across different databases and therefore enable products from multiple companies to work together. The integration of CORBA and database technologies enables clients to access CORBA objects that are stored in the persistent storage.

In object-oriented systems, objects accessed by the client must be loaded in the client address space of the client machine. In CORBA the client request are processed on the remote object, and the result is returned back. Furthermore, in relational database management systems, clients make a query call to the database, thus exposing the relational scheme to the client. In CORBA, objects export an interface, defined in IDL, to their clients. The choice of the database management system is not revealed to the clients.

Wireless middleware is a relatively new type of middleware that has the same function as traditional middleware but it is specifically designed for wireless networks. Typically it has architectural features such as home and foreign agent functions to support the use of mobile IP protocols so that wireless users can move across different subnets. However, in the context of the SCA, middleware is used for communication between different computing elements and wireless middleware is not particularly relevant for this purpose.

6.2.5 POSIX and RTOS

Fundamentally the SCA requires that the system is implemented on a machine that uses a POSIX-compliant OS with real-time embedded capabilities such as preemptive multitasking. Note that waveforms are not fully independent of the OS; waveforms will need limited access to the OS in order to use threads and processes.

POSIX stands for Portable Operating System Interface for computing environments. POSIX is a standard environment to enable portability of applications software. POSIX support assures code portability between systems and is increasingly mandated for commercial applications and government contracts. For instance, the USA's Joint Technical Architecture—Army (JTA-A) standards set specifies that conformance to the POSIX specification is critical to support software interoperability. The portability of a software application is achieved through the specification of a set of services that every POSIX conforming application can expect to exist on a conforming platform. POSIX has started as a standard enabling portability across UNIX environments but is not limited to UNIX. There are other OS and RTOS environments which support POSIX fully or partially. POSIX allows for significant reduction in cost during porting code. Real-time embedded developers are usually looking for POSIX conformance. POSIX conformance means that the POSIX.1 standard is supported in its entirety. The standard also defines extensions for real time which are interesting for SDR development. The POSIX.1 standard specifies application programming interfaces (APIs) at the source level, and is about source code portability. It is neither a code implementation nor an operating system, but a stable definition of a programming interface that those systems supporting the specification guarantee to provide to the application programmer.

The real time operating systems are categorized as hard real time and soft real time OS based on how deterministic they are as far as delivering functionality with respect to time. In many cases soft real time operating systems like Linux are sufficient to do the job since they provide rich interfaces and a large set of development tools as well as popularity among developers. For a RTOS the following characteristics are valid:

- They use much less memory compared to regular OS
- They have flexible schedulers which support fine control of parallel tasks
- They have deterministic response in controlling the software tasks
- They are not very bulky and can be customized to fit into smaller CPUs with not very high demands for processing power
- Most of the modern RTOS offerings conform to the POSIX standard
- They provide full networking capabilities and middleware solutions

Most RTOSes on the market are offered for general purpose processors (GPP) since those chips are used extensively for variety of products. There are some vendors who offer RTOS for DSP chips. RTOS solutions can be offered for FPGAs which have a CPU core bundled within the

FPGA. Ideally with careful planning all chips in an SDR solution can use an RTOS.

In a wireless system the RTOS will reside in the controller. Such an RTOS is a multi-tasking environment. Since in this environment there are multiplesimultaneous tasks of execution, communication among them is very important. The RTOS supports many forms of Inter Process Communication (IPC) but those are mainly used to communicate within the same memory space. In order to communicate between distributed CPUs a different approach base on TCP/IP communication is needed. Such communication can be accomplished with CORBA. The controller must contain the OS and the ORB.

Suppose that two functional blocks must be connected and one is implemented on an FPGA and the other on a DSP. These blocks need device drivers for the given RTOS. In addition to the device drivers, a CORBA wrapper (a.k.a. CORBA proxy) around the driver is necessary to provide an entry point for that device into the RTOS's environment. The proxy allows the hardware to become part of the CF, i.e. to appear as a software component.

6.2.6 CPUs and Software Development

As mentioned in the Introduction, modern high-performance software-defined radios require several types of CPUs, such as GPPs, DSPs and FPGAs. These different hardware technologies are required in order to satisfy the timing and functionality requirements of SDR projects. DSP's are typically used to implement low speed signal processing portions of the baseband subsystem. FPGAs implement higher-speed signal processing portions. Applications are implemented on GPPs. The combination of FPGAs, DSPs, and General Purpose Processors (GPPs) requires a fairly diverse set of software tools and programming languages.

Today most DSP code is written in C often using code generators. It is then tested for ability to meet critical timing constraints. Code modules which are on the critical timing path and which don't meet requirements are then hand coded in C. If they still don't meet the timing requirements, they are hand coded in Assembly language. If they still don't meet timing requirements, they are hand coded in machine language. If they still don't meet timing requirements, that functionality is moved to the high-speed signal processing portion.

C++ is being used because it enforces the OOD technique discussed earlier. However, if C++ is used as the starting point, then the timing optimization process outlined above may have one additional step. Developers have come to the conclusion that they can write more efficient code by writing in C and keeping the Object Oriented Model in their head. C++ programs are typically written in human-readable form and then compiled to produce an executable specific for a given processor. To produce an executable for a different processor, the code must be recompiled. If the processors have different architectures, different memory organization, etc., this recompilation may not be possible. Even if possible, it usually results in an executable that has very different performance.

As the software developer moves down this abstraction ladder from pure algorithmic (C) to hardware specific (Machine Code), improvements in performance come at the expense of detailed knowledge of the underlying hardware. This detailed knowledge is very specific to particular semiconductor products and is neither easily transferable nor generalizable.

Insulating the programmer from the hardware has several attractive benefits. It can speed the development and can also allow portability. Java with its JVM (Java Virtual Machine) provides this insulation. Unfortunately, although dramatic improvements in Java run time performance have been made, the ever increasing processing demands of the baseband subsystem exceed Java's real time performance ability. Java tries to combine the advantages of platform independence and compilation. Java programs are first compiled into a set of byte codes, which are then interpreted over the Java Virtual Machine (JVM). Java programs are compiled only once and interpreted every single time it is run. The use of the JVM allows the software application to be used on multiple devices. This allows interoperability over a wide variety of H/W and the use of third party S/W. Java is seeing a strong base of use in wireless applications running on the Application Processing Subsystem. It has some advantages for SDR systems. It is a more dynamic and adaptable language than C++, and is also well-suited to networking.

However SDR systems place strict timing constraints, which the JVM may be unable to meet. Current technology development is trying to overcome this disadvantage and create a virtual machine that offers real-time performance. The requirements towards any virtual machine are: direct access to system devices, multi-threading capability (multiple buses, registers in both shared and distributed memory), variable-bit precision fixed and floating point arithmetic, and adequate real-time performance. There are also optimized JVM which run directly on the hardware with no operating system. This approach speeds up further the execution of Java code.

The first SDR's were implemented with FPGA's (Field Programmable Gate Arrays). FPGA code is often generated based on input in C or VHDL (Very High Level Hardware Definition Language), Verilog or even System C. In some cases, a tool that takes C as an input and produces VHDL, is used and then the VHDL code is passed to a code generator that produces FPGA code. FPGA code is typically a combination of device functionality specification and device layout. Some tools can go directly from a high level description language to FPGA code. Hand optimization may be used to improve performance. Because of the nature of FPGA's, hand coding takes on many of the characteristics of hardware design.

The use of code generators is growing. These generators can start with C as the basic input. However, a new class of generators is becoming prominent. In these input is in a high level description language such as UML (Universal Language) and the tool outputs C or VHDL. Finally, there are tool chains where the designer inputs high level description language and specifies which portions will be implemented in DSP and which ones in FPGA. The tool outputs DSP code (typically in C) and FPGA code (in the form dependent on the types of FPGA). There are also tools chains which operate in a similar fashion but incorporate the SCA.

6.3 The Operating Environment and the Core Framework

The Operating Environment (OE) contains the bulk of the SCA components. It includes the Core Framework, the OS and middleware interfaces, and the Domain Profile. The OE includes Name service, Event service, Log service, File, File system, file manager, Testable object, Resource and Resource Factory, and Port [8].

The Core Framework (CF) is the essential set of open interfaces and services at the application layer that provide an abstraction of the underlying system software and hardware. The CF consists of:

- **Base Application Interfaces**: *Component Identifier*, *Controllable Component*, *Port Accessor*, *Life Cycle*, *Testable Object*, *Property Set,* and *Resource*, which provide the management and control interfaces for all system software components.
- **Base Device Interfaces:** *Capacity Manager, Device Attributes, Device, Manageable Component, Loadable Device, Loadable Object, Executable Device*, and *Aggregate Device*, which allows the management and control of hardware devices within the system through their software interface,
- **Framework Control Interfaces:** Application, Application Deployment Data, Application Factory, Component Registry, Domain Installation, Domain Manager, Domain Manager Attributes, Event Channel Registry,

Full Component Registry, Manager Registry, Full Manager Registry, and Device Manager, which control the instantiation, management, and destruction/removal of software from the system,

- **Framework Services Interfaces**: *File, File System, Component Factory, Component Manager, and File Manager*, that provide additional support functions and services.

The relationship between the CF, the SCA applications, and the hardware platform is shown in Figure 6.13 [2]. An application consists of multiple software components that operate over a distributed system. Communications between the application and the SCA Devices is accomplished through the CORBA middleware. An application may access only mandatory OS functionality. The reason the SCA includes POSIX interfaces is to provide portability for those waveforms that use OS services. This is mostly relevant for those software components implemented in a GPP.

A UML illustration of the SCA is shown in Figure 6.14. Software components are partitioned into three groups: Radio Management, Devices and Applications. Every software component has ports, configuration parameters, requires resources and provides capabilities. The implementation of a component is hidden from the interfaces.

A *Resource* has a certain combination of input, output, and control ports. Resource will most likely create an instance of a Port within the Resource.

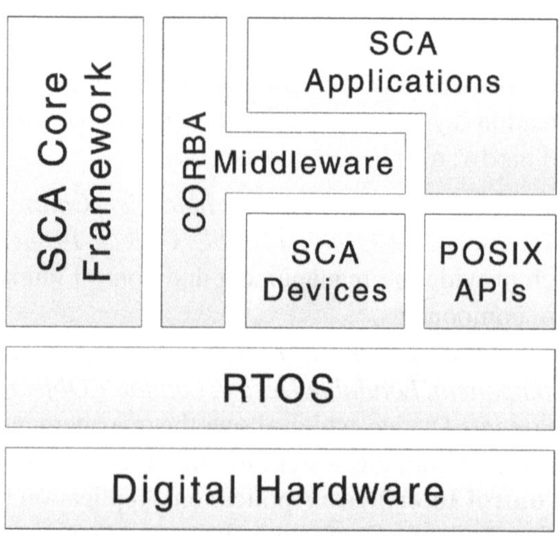

Figure 6.13 SCA-based system architecture.

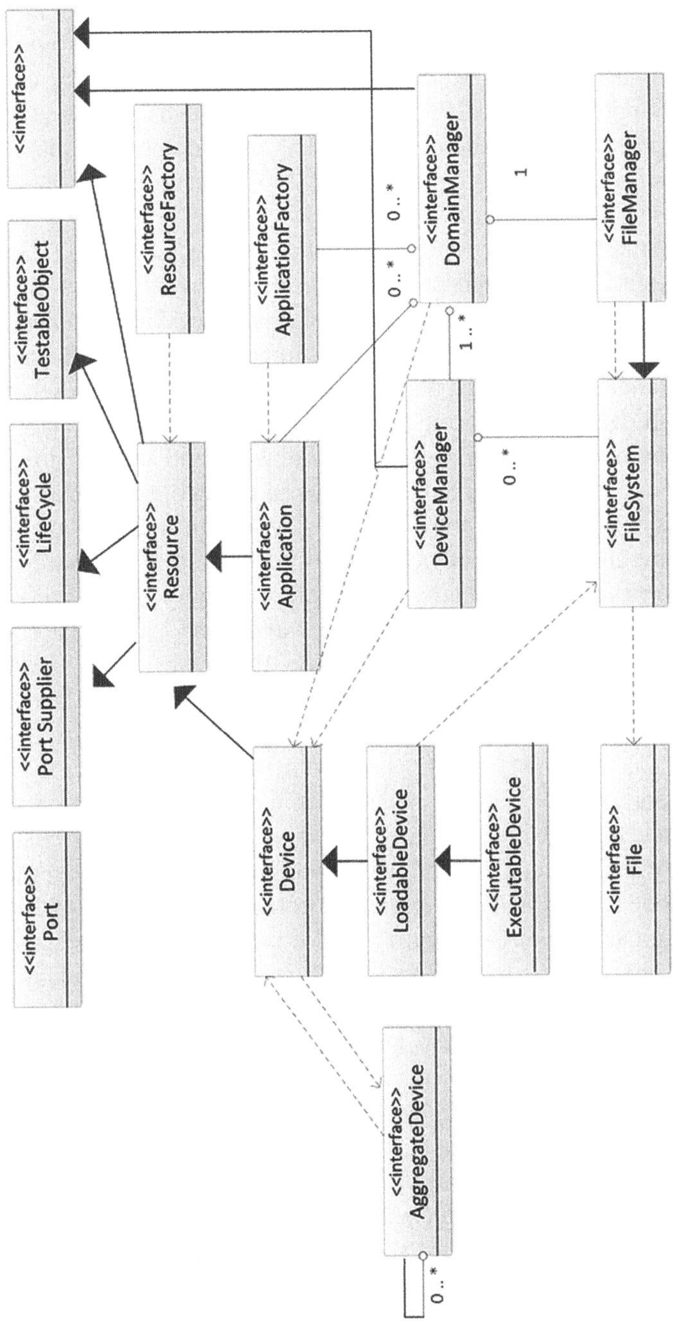

Figure 6.14 UML model of the SCA.

Port contains two member methods – connect Port() and disconnect Port(). Note that Port is a component-specific implementation. To use a *Resource* in an application, it must be deployed onto a *Device.* A *Resource* has capability (OS, processor) and capacity (MIPS, memory) requirements a *Device* must meet. For example, a simple resource is an FFT. A more complex resource is audio coding, etc. Applications can be started, stopped, configured, queried, tested, etc.

There is a hierarchy of three types of devices - Device, Loadable Device, Executable Device. The software components that provide access to the system hardware resources are referred to as SCA "Devices" that implement the base Device interfaces. The Device class inherits from the Resource class. It has two member methods: allocate Capacity() to set aside a specific HW capacity and deallocate Capacity(). Loadable Device class extends the needs of the Device class through its ability to load and unload images. Devices can be started, stopped, configured, queried, tested, etc. *Devices* are software proxies (or interfaces) for hardware devices. In this way the line between hardware and software devices is blurred. Note that non-hardware (software-only) resources provided by the system for use by applications are generically referred to as "services". When a waveform is instantiated the existence of all the devices required to support the waveform is verified and the capacity is allocated. Every processor is an *SCA Device* and the same *load* operation is used, regardless of whether it is a FPGA, DSP, or GPP.

A radio is composed of many nodes. For example, a board can be considered a radio node. More generally, any device capable of interfacing through CORBA can be considered a node. A radio node has *Devices* and therefore each node runs a *Device Manager* that contains complete knowledge of the devices and services. A system may have multiple device managers. Each *Device Manager* must register to their *Domain Manager*, making its components known to the radio. Device Manager does not only manage devices. It also creates and maintains a File System with Platform-specific storage.

Radio Management is performed by four components: *Domain Manager, Application Factory, Application, and Device Manager. Domain Manage* is the core of the SCA system. It is used by UI to control/monitor the radio. The Domain Manager maintains a comprehensive list of all existing applications (and their resources), all file systems, and all device managers. *Application Factory* is used to instantiate an application and to provide an *Application* component. *Application* is used to control a deployed application. In addition to *Domain Manager* and *Device Manager*, additional software components that are required by the SCA are *File Manager* and *File System* interfaces (Figure 6.15). Any software component can access a file anywhere in the radio platform. SCA radios are not allowed to use the file system provided by the OS.

The CF is generally split into two sections, the red (secure) and black (non-secure) sections. The dual connection with black and red buses is only relevant for defense systems and is not expected to be used for commercial systems. Different security mechanisms are employed for commercial systems.

6.3.1 The Domain Profile

Every SCA-compliant radio can be considered to be a domain. The domain is the collection of devices, software components, and applications that reside within the system. Now it becomes intuitively clear that the Domain Profile is a hierarchical collection of descriptor XML files. The SCA Domain Profile elements identify the capabilities, properties, inter-dependencies, and location of the hardware devices and software components that make up an SCA-compliant system.

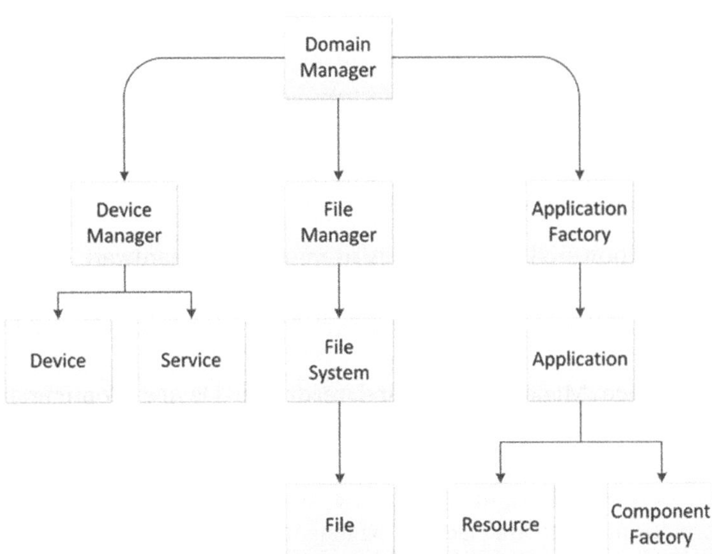

Figure 6.15 SCA Management hierarchy.

Since applications are composed of multiple software components a Software Assembly Descriptor (SAD) file is defined to determine the composition and configuration of the application. The SAD is like a schematic diagram for "wiring" the waveform. The SAD references all software elements needed for this application, defines required connections between application components (connections ports/interfaces), defines needed connections to devices and services, provides additional information on how to locate the needed devices and services, defines any co-location (deployment) dependencies, and identifies component(s) within the application as the assembly controller [6].

Each software element in the system is described by a Software Package Descriptor (SPD) and a Software Component Descriptor (SCD) file. The SPD provides identification of the software (title, author, etc.) as well as the name of the code file (executable, library, or driver), implementation details (language, OS, etc.), configuration and initialization properties (contained in a Properties File), dependencies to other SPDs and devices, and a reference to a Software Component Descriptor. The Software Component Descriptor (SCD) defines interfaces supported and used by a specific component. There is also *Domain Manager* Configuration descriptor XML file.

In addition, XML files are used to describe the radio hardware. Note that the Domain Profile may not be static. When a new hardware or a new waveform is installed, the Domain Profile will change to support hardware and software modularity. The Application Factory verifies that the requirements of the application match the hardware capabilities.

6.3.2 Power-up Scenario

When the radio is powered-up the goal is to initialize all hardware elements and to load the elements of the Core Framework. The power-up sequence is starting Device Manager, Domain Manager, Event Channel, File System, and Log.

First the Device Manager starts and reads the Device Configuration Descriptor and launches all nodes. (A 'node' is a physical hardware component). The Device Manager starts the Domain Manager and initiates also a number of services and applications: File system, Log service to record system information. Once the Device Manager has processed all devices appropriately, it informs the Domain Manager that its work is complete. More Device Managers may be started, but each subsequent one will not start a

Domain Manager nor the event, file, and log services, since there can be only one of these per system.

The Application Factory is used to create the waveform. It reads and parses the XML files that describe the hardware and software components necessary to deploy a waveform on a system. Then it identifies the resources required, and creates whatever resources are needed that have not already been created. The software components are not yet loaded onto the corresponding hardware components. Properties of the application are set to default or initial values, specified in the XML files. These values may be changed prior to actually starting the application or during the operation of the waveform.

When an application is instantiated, capacity on the hardware components is allocated and the software components are loaded. Then the Application Factory also connects all Ports, starts the waveform software on the required hardware, and informs the Domain Manager that it has successfully created the waveform. A name reference is obtained from the naming service. This name is connected to the appropriate input and output.

6.4 SCA Next

The first versions of the SCA were found to be more suitable for larger, multi-channel radios. Radio management, for example the power-up and instantiating a waveform procedures described above are slow. The default communication protocol for CORBA is TCP/IP, which can lead to latency problems. CORBA also needs to allocate and de-allocate memory dynamically, which must be managed carefully to avoid memory fragmentation issues and time jitter.

The recently released SCA Next [4] is developed to address these concerns. To provide scalability, SCA Next includes three SCA profiles: Full, Medium, and Lightweight. The Lightweight profile is targeted towards environments with limited computing support. It does not support hardware modularity and does not offer interfaces through which hardware components can register and unregister with Device and Domain Managers. In the Lightweight profile Resource, Device, Device Manager, and Domain Manager interfaces are not fully implemented. The benefit from the elimination of interfaces includes smaller executable size. Under the Lightweight profile on power-up the radio begins executing a single waveform with minimum reconfiguration. This is intended to satisfy the requirements of low-power and single channel radios.

The Full Profile supports all interfaces for registering and unregistering components to Domain and Device managers. It offers plug-and-play capability and supports complete hardware modularity. The Medium Profile is suited to radio platforms where the hardware architecture is static. It supports registration, but not un-registration of components.

These profiles have a hierarchical structure: the Full Profile is a superset of the Medium Profile, which is a superset of the Lightweight profile.

The OE has a similar hierarchical structure, with several overlays of functionality. These overlays include

- Support for dynamic installation and uninstallation of applications
- Support for CORBA
- Support for event channel and event service
- Supports the log service
- Deployment of applications onto platform channels

The way scalability is achieved is via the use of optional inheritance. For example, using optional inheritance, *Executable Device* and *Loadable device*, which is different from the normal inheritance rule illustrated in Figure 6.16. no longer inherit from *Device*. Components always implement a single interface, but what this interface inherits from is decided using compiler directives resolved at IDL compile time (Figure 6.17). This not only provides scalability, but is backwards compatible with the SCA 2.2.2 (the version preceding SCA Next). The disadvantage is that it violates the UML inheritance rule. SCA Next deals not with single IDL file, but a collection of IDL files. Note that to avoid problems at link time, IDL that has been translated differently must reside in different address spaces.

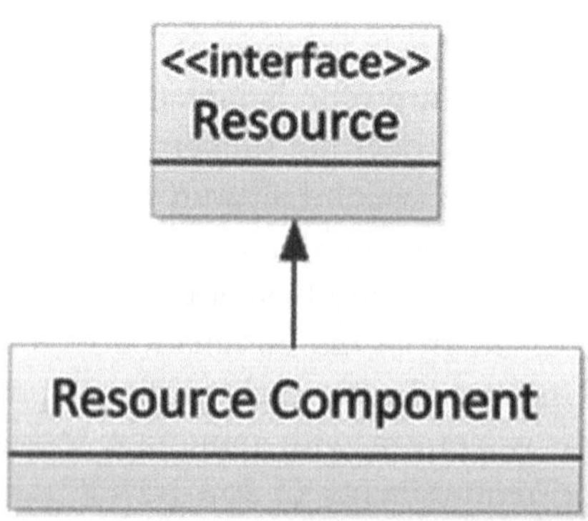

Figure 6.16 Inheritance in SCA 2.2.2.

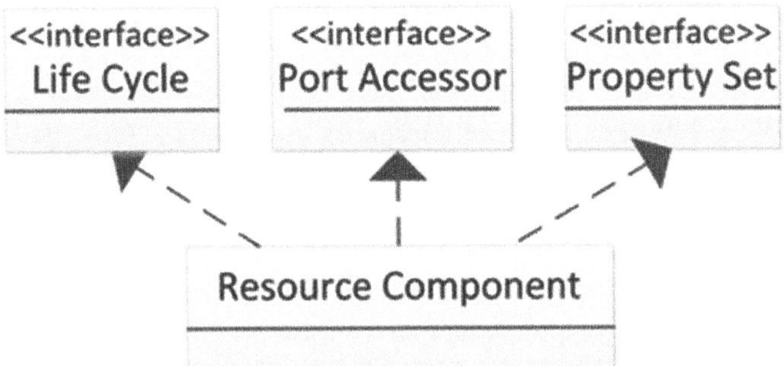

Figure 6.17 Optional inheritance in SCA Next.

Another change under SCA Next is that CORBA is no longer required, but optional. To accomplish this middleware-independent APIs are defined, although they are specified through IDL. There is no Naming service and the Application Factory maintains a registry of all components. When a waveform is instantiated, the Application Factory will send the load and execute commands to the processor. The waveform once started will register with the component registry and then the application factory will connect it with other software components by providing them with the object references necessary to communicate.

The SCA Next reduces the cost of platform development, waveform development, and the SCA certification. It also provides better performance, allowing products with lower Size, Weight, and Power (SWAP), and better performance, such as shorter boot-up time period and improved real-time performance.

6.5 SCA Specifics for DSPs and FPGAs

The easiest approach to build an SCA-compatible radio is to use a GPP. GPPs have normally RTOS solutions available, a complete implementation of CORBA, and other software libraries. The reason we can't solve all SDR problems with GPPs is because GPPs are very slow for many of the tasks interfacing with the analog world in real time.

A waveform may be realized using a GPP, DSP or an FPGA or any combination of the three depending on the real time requirements and complexity of the algorithms involved. The software is running on different hardware entities in parallel which brings challenges in the timing and synchronization of the solution. This brings some challenges because the implementation is a mix of VHDL for the FPGA, C language for GPP and C or assembly for the DSP. This development process is difficult. Especially challenging is the validation of the solution as far as functionality and time is concerned, even when special tools are used.

DSPs are specialized chips which can be used for real time processing of digital filters. They have special hardware enhancements which make the execution and development of digital signal processing algorithms very efficient. DSPs have evolved alongside GPPs following two trends. One trend is to provide more processing power for signal processing and the other trend is to provide more standard GPP functionality. Some of the DSP offerings today are very versatile and may combine the features of a GPP and DSP in one core. Similarly the development tools for DSPs are becoming more "user friendly" and easier to use.

Today's FPGAs are fast and can be used for the most time demanding tasks in the project. The development is done in VHDL or Verilog which are hardware design languages.

In general VHDL or Verilog development is a more difficult endeavor compared to development for GPPs and DSPs.

When using a standard FPGA the interface to the device is through low level registers. The SCA requires each device to be represented as an SCA device. Then the Device Manager as specified by the SCA can handle those devices as similar entities. For this to happen though each device has to provide flexible software interfaces which isolates the low level details of the device from the application software. This can be accomplished by using XML files which describe the mapping of the high level interface to the low level specifics of the device.

DSPs and FPGAs normally do not have an RTOS, neither middleware support, which makes them different environments for SCA development.

Figure 6.18 shows one approach to organize the communications between processing entities which do not have native CORBA implementations. In this case the GPP with the RTOS running on it provides CORBA software proxies. These proxies provide the basic CORBA functionality and do the translation between the CORBA and non-CORBA world of the DSP and FPGA chips. This solution is general and solves the problem, but can be associated with considerable latency.

The Modem Hardware Abstraction Layer (MHAL) device, shown in Figure 6.19, is developed for are processors and bus technologies without off-the-shelf CORBA support and allows waveform components hosted on GPPs,

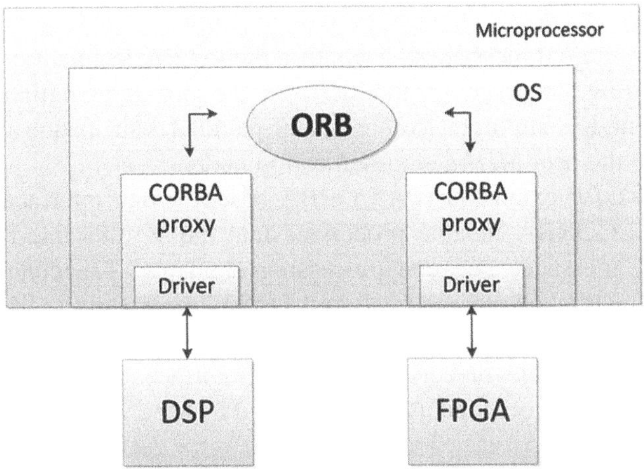

Figure 6.18 Communication among entities that do not have CORBA implementations.

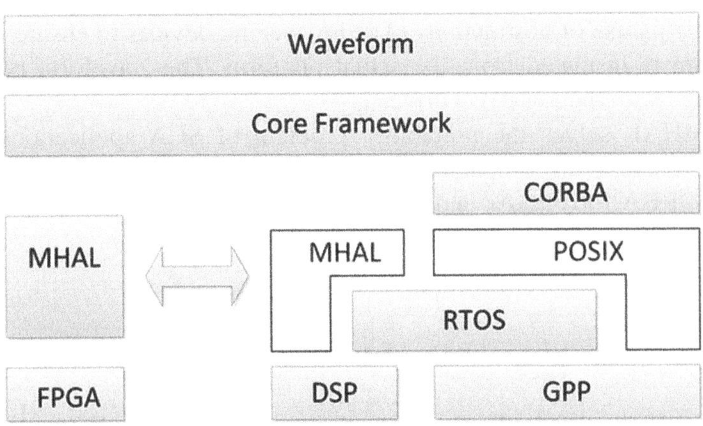

Figure 6.19 The SCA CF over FPGA, DSP, and a GPP.

DSPs, and FPGAs to communicate. *MHAL* specifies a consistent method for routing messages between each computational element. It achieves this through the definition of *MHAL* interfaces and structures. It is an example of a component-level adapter and abstracts the waveform application from the specifics of the hardware [7].

If another off-chip component is used, that component should be be abstracted using MHAL. Cases, where performance requires the waveform to be tightly coupled to the platform, should be avoided, or at least minimized.

The structure of an *MHAL* message is very simple. It consists of an In-Use (IU) bit for internal message flow control, the logical destination, shown in Figure 6.20. (e.g. end user) for the message, the message length, and the payload to be sent. It is important to note that the payload is an opaque message that is passed to the peer *MHAL* communication node.

The *MHAL GPP* extension is a CORBA-based device interface documented in IDL. CORBA-capable processors may utilize this interface for sending *MHAL* messages between processing elements. The concept for defining an infrastructure on the DSP and FPGA processors is similar to that of the *MHAL GPP*. Because the DSP environment does not readily support dynamic linkable objects, the *MHAL* interface on the DSP is a library of standardized components that are linked into the waveform code at build time.

MHAL allows waveform software to be ported seamlessly from another computational element to another (Figure 6.21). The *SCA Device* is an abstraction which sits on top of the MHAL layer. One of the characteristics of the SCA is its use of abstractions of radio-specific devices to ensure that the waveform is insulated from the actual platform. The waveform never communicates with the low-level drivers.

While MHAL solves the portability problem of SCA applications, it would be beneficial if all computational elements (GPP, DSP, FPGA) support CORBA. CORBA for FPGAs and DSPs is indeed becoming available and makes connecting different processing environments more straightforward.

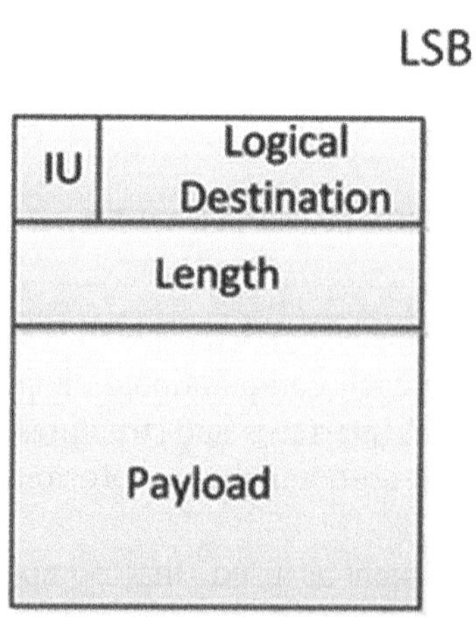

Figure 6.20 MHAL message structure.

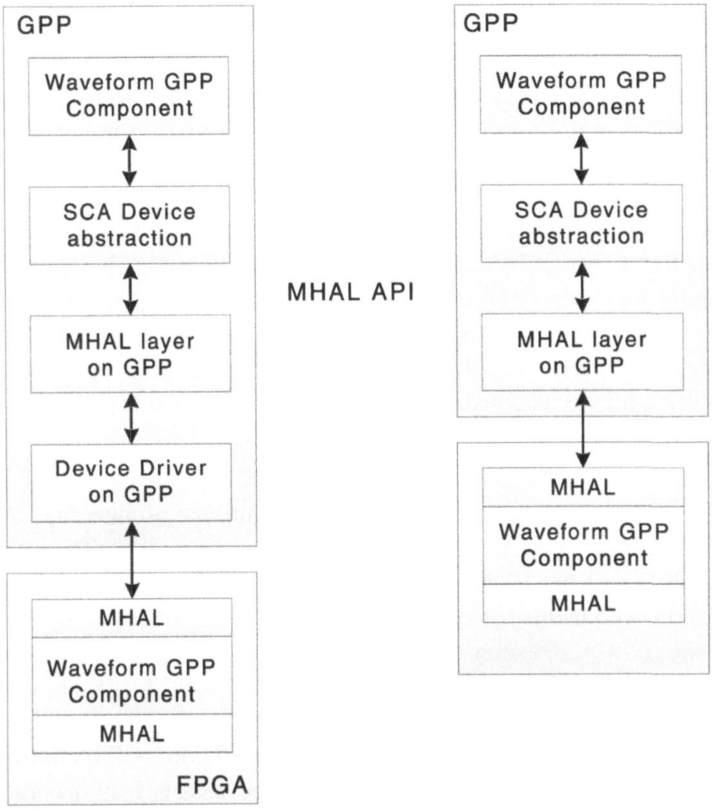

Figure 6.21 Seamless porting of waveform software from FPGA to DSP.

Newer DSP and FPGA chips provide some GPP functionality and a standard CPU core which can behave very much like a standard GPP.

6.6 Design Principles

6.6.1 General Software Rules

A high-level language is strongly recommended, since it clearly facilitates code portability. Furthermore, platform-specific and compiler-specific features of high-level languages should be avoided. If there are timing requirements that cannot be met using high-level language, assembly is required. Such cases should be rare, and very well documented. Furthermore, features of C/C++ that are not portable to other platforms and different compilers should be avoided.

It is possible to use legacy software in an SCA-compliant system. This can be done using a wrapper that provides SCA interfaces [5].

There are software tools that considerably facilitate the development of an SCA-compliant system. Spectra CX from Prism Tech is a model-driven development tool that simplifies, accelerates, and validates a significant part of the Software Communications Architecture (SCA) development process. Spectra CX validates SCA compliance at the architectural and unit test level, and generates correct-by-construction SCA compliant artefacts, such as: XML descriptor files, compliance test reports, and validation documentation. Spectra CX enables SCA and non-SCA software aspects to be developed together, integrated early, and thoroughly tested. Spectra CX also reduces development risk due to its consistent model-based approach [3].

6.6.2 Hardware Architecture Requirements

Note that the SCA does not require any specific hardware architecture. Any mix of devices can be used in order to fulfill the requirements. This stems from the flexibility of the SCA which can be deployed easily over diverse and distributed components. Still when selecting the hardware components their processing power measured in MIPS is a guiding factor for the types of algorithms that need to be executed. For example a 50,000 MIPS algorithm can't run on a GPP with a 1000 MIPS. When using an FPGA to realize an algorithm in VHDL it is not clear how much the speed gain will be until the design is finalized and timing analysis is performed. That is why it is really hard to place a number in MIPS or Dhrystones regarding an FPGA block which is used to perform certain function. To complicate things further portions of the solution can be done in the FPGA and portions in a DSP or GPP which some modern FPGAs come bundled with. One safer and cheaper approach is to use simulators available from the vendors of the chips before the actual prototype is ready and the project is further in the product cycle.

6.6.3 System Architecture Overview

The system architecture can determine many characteristics of the entire product. Some of the decision that are made at this level can affect the following characteristics of the product:

- Power consumption. Careful selection of algorithms has an impact on the power consumption of the hardware. Careful selection of data communication can also impact consumption. Similarly the software

architecture can have an impact on consumption too. For example if there are periods of no communication and the software waits for an event by keeping a background process running this can save energy.
- Responsiveness and timing. Choices like usage of integer arithmetic or usage of lookup tables can make a big impact on the speed of the solution. Similarly specific techniques for memory usage in the software like static allocation of resources can speed things up too.
- Reliability. Any system is as reliable as the weakest link. Thus creating a reliable system with so many hardware and software components is a challenge. Considering parallel software and hardware solutions within reason is always worth thinking about. Constructing recovery schemes in software is also worth the investment in time. Avoidance of any dangerous software techniques like dynamic memory allocation and dynamic process creation as well as better error handling can lead to improvements in reliability of the system.

6.6.4 Interface Organization

In addition to the recommended SCA framework which delivers the basic means for real time communication via middleware the designers of SDR systems can increase code reuse and modularity of their designs by defining appropriate software interfaces. Those interfaces can define the basic services each module can deliver for certain line of products. The interfaces are realized as messages which are sent and processed at the application layer. The definition of those interfaces decouples entire subsystems which can then be replaced with newer CPUs, RTOSes and even middleware. Thus the application part of the waveform generation becomes portable and this leads to better code reuse and shorter time to market. By devising a common message format the parsing of messages can be done with the same code across different devices. This further eliminates duplication of code and functionality. Such interfaces makepossible the distributed development of a solution by geographically distant teams.

6.7 Summary and Future Directions

In summary, the SCA specifies

- A standard set of interfaces for logical devices
- A common file system interface and services

- A logging and event service
- A common interface protocol (CORBA)
- A standard method of locating components (Naming Service)
- A method for connecting components
- A set of components and interfaces that configures, installs, controls, and monitors applications and devices
- Domain Manager, Application Factory, Device Manager
- A standard method for specifying components, resources, and connections to make a waveform (using XML files)
- A high-level guide for API development and security management

The SCA is a specification standard and a component-based software framework for SDR. Components are typically concerning software entities but may refer to hardware. The goal of SCA is to provide rules and behaviors which make seamless integration of those components into a complete solution. Using component based development allows engineers to concentrate on the application and thus achieve faster time to market and better software reuse.

SCA is suitable for systems built using heterogeneous types of digital hardware and is a platform independent framework, supports multiple operating systems, multiple processor types, and many external devices. The ultimate benefits are software and hardware portability, design and code re-use, and reduced development cost.

Currently the market for the SCA is within the defense-related industry. There have been attempts to make the SCA a commercial standard. However, hardware/software portability for reconfigurable embedded systems with heterogeneous types of processors comes ultimately at a cost. The benefits of design and code re-use may not go to the companies that adopt these design principles. The SCA 2.2.2 generally has not been accepted by the commercial wireless industry, although SDR technology is being integrated into commercial cellular base-stations. Currently, the base-station vendors are maintaining a proprietary and confidential approach to how they integrate SDR. For commercial cellular handsets there are a number of competing technical approaches attempting to provide a cost effective, power efficient SDR solution. The solutions are ad hoc and do not involve the SCA and any of its components such as the Core Framework. It is likely that in the commercial SDR market, a path similar to that of early software development will be followed. Initial solutions will continue to be ad hoc and proprietary until third party software developers will begin to appear.

The use of the SCA outside of the defense industry will most likely start by the public safety radio industry. System developers of public safety communication systems have come to conclusion that certain parts of the SCA can be useful. The lightweight profile within the SCA Next is an interesting possibility for both public safety and commercial systems. It can satisfy the speed and size requirements of not only high-end systems but also to smaller footprint systems with less power in their hardware resources.

The SCA is component-based, domain- and platform-independent and not bound to one specific operating environment. It can be used not just in radios, but in all embedded systems built using heterogeneous types of digital hardware which can benefit from its features. Future areas of application include avionics, aerospace, vehicles, etc.

References

[1] J. Bickle, "Next-generation SCA operation environments," Proc. SDR'06 Tech. Conf., 2006.

[2] Communications Research Center, Introduction to the SCA, Wireless Innovation Forum, May 2011.

[3] http://www.prismtech.com/spectra/products/spectra-cx.

[4] Software Communication Architecture Specification, version Next, Joint Program Executive Office, 2010.

[5] Joint Tactical Radio System Network Enterprise Domain, Test and Evaluation, Waveform Portability Guidelines, Dec. 2009.

[6] F. Levesque, S. Bernier, "Interconnecting SCA Applications," Proc. SDR'07 Tech. Conf., Denver, CO, 2007.

[7] C. Magsombol, C. Jimenez, and D. R. Stephens, "Joint Tactical Radio System – Application Programming Interfaces",Proc. IEEE Military Communications Conf., 2007.

[8] K. Richardson, C. Jimenez, D. R. Stephens, "Evolution of the Software Communication Architecture Standard," Proc. IEEE Military Communications Conf., 2009.

7

An Approach to Resource Management in Future Internet

Oleg Asenov
Department of Informatics, University of Veliko Tarnovo, Bulgaria

Vladimir Poulkov
Technical University of Sofia, Bulgaria
E-mail: Vladimir Poulkov<vkp@tu-sofia.bg>

Albena Mihovska, Ramjee Prasad
Center for TeleInfrastruktur (CTIF), Aalborg University,
Aalborg, Denmark

7.1 Introduction

The Internet has been existing for more than two decades now and has evolved from a network of routers and domains to support a range of basic functions (e.g., file sharing, remote login) into a complex pervasive network of networks where information is routed in a dynamic and highly distributed fashion to a myriad of end-devices, most of which highly personalized and supporting business, information search, or social life. The ubiquity of the Internet and the unpredictability of its operations have given rise to new research problems, an important one related to how to enable reliability and quality of service (QoS) despite the constraints imposed by the complexity and dynamical behaviour of the plethora of Internet links, which are in essence connections between various autonomous systems. The Internet is constantly increasing its size and dynamics, due to the huge advances of the wireless communication technologies and hardware that have penetrated all spheres of every-day life. The demand for very high wireless data rates imposes requirements of similarly matching responses from the supporting Internet infrastructure. The future Internet is evolving towards even more complexity to support services and applications originating from large-scale networks of

Vladimir Poulkov and Ramjee Prasad (Eds.), Resource Management in Future Internet, 185–210.

information collecting and sending devices. Many of the enabling Internet operation technologies, such as routing, resource and access protocols, which depend on a knowledge of the network topology, are faced with scaling limitations.

A particular challenge is to enable timely and context-aware resource provisioning for dynamic services and application requests. Such handlings should also be kept transparent for the users and underlying connections. This Chapter proposes a supporting transition architecture from internet based multimedia towards multimedia-based Internet where QoS is provided by optimisations to protocols and algorithms for proactive resource allocation and management.

This Chapter is organized as follows. Section II focuses on the challenge of QoS provision for the plethora of Internet services and applications. Key techniques, such as Access Node Control Protocol (ANCP) and Multiprotocol Label Switching (MPLS) are also described. Section III focuses on the key aspects of the transition from Internet-based multimedia to multimedia based Internet. An idea of a transition architecture to Internet is also presented together with the practical problems to be solved. Section IV summarizes the key points of discussion and concludes the Chapter.

7.2 Resource Management and Quality of Service

The key problem with Internet Resources Management (IRM) is related to the necessity of providing a specific QoS for the different types of services and applications, such as access to web pages, exchange of short data communications and files, telephone and video-telephone calls and conferences, access to on-line radio and television, etc. The type and frequency of the users' requests and the traffic generated are random and difficult to be estimated, because they are a function of geographic, demographic, social and economic factors. At the same time, the Internet network architecture should provide the possibility of a Resource Management (RM) practically independent of the type and frequency of users' requests while keeping the QoS relatively constant at an acceptable level. On the other hand we will face substantial difficulties if we try to define the term Internet resource, and that is because practically every "IP-visible" object on the global network can be considered a resource. According to the Internet Engineering Task Force (IETF) experts, two key resource entities can be determined for the Internet access architecture [1] as listed below and the management of which provides the actual users sessions the required QoS, as shown in **Figure 7.1.**

- **Access Node (AN).** Network device, usually located at a service provider central office or street cabinet, that terminates the access loop connections from subscribers. In the case of Digital Subscriber Lines (DSL), this is often referred to as a Digital Subscriber Line Access Multiplexer (DSLAM). In Passive Optical Networks (PON), this is usually comprised of an Optical Network Termination (ONT) / Optical Network Unit (ONU) and an Optical Line Termination (OLT);.
- **Network Access Server (NAS).** Network device which aggregates multiplexed Subscriber traffic from a number of Access Nodes. The NAS plays a central role in per-subscriber policy enforcement and QoS. This is often referred to as a Broadband Network Gateway (BNG) or Broadband Remote Access Server (BRAS). A detailed definition of the NAS is given in RFC2881 [1].

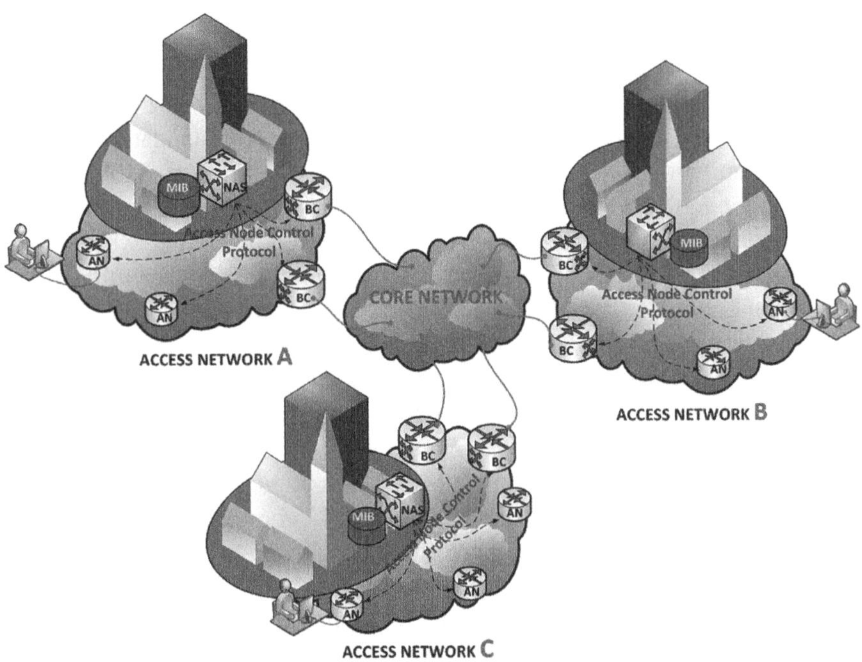

Figure 7.1 Internet access architecture.

The AN and NAC together with the communication lines that connect them form the so called Access Networks, giving the users access to resources and applications on the Internet.

According to the same IETF group specialists IRM and requested QoS assurance is effected on basis of the Access Node Control Protocol (ANCP) which during the last three years is applied as a basic functionality of the network equipment developed and offered on the market by the leading manufacturers in this segment, like Cisco, Juniper and Huawei [2–4].

The ANCP idea is based on the technological opportunity to monitor and store by means of IP protocol based Simple Network Management Protocol (SMNP) the condition of every AN and Border Controller (BC) of the specific Access Network (Access Network A, B, C) in the so called Management Information Base (MIB II) and decisions to be "made" by NAS on configuration changes depending on the current condition of service requests [5]. When discussing Internet resources management it is important to note that there are more than hundred of protocol specifications in the form of RFC's recommended by IETF [6]. Contemporary architectural solutions for Access Networks management are based on the protocol specification - MPLS (Multiprotocol Label Switching). MPLS key role in modern Internet architecture development in general is confirmed also by the fact that IETF has created its own working group which develops recommendations on MPLS implementation [7]. The MPLS working group is responsible for standardizing technology for label switching and for the implementation off label-switched paths over packet based link-level technologies. The working group's responsibilities include procedures and protocols for the distribution of labels between Label Switching Routers (LSRs), MPLS packet encapsulation, and for Operation, Administration, and Maintenance (OAM) (including the necessary management objects expressed as MIB modules or using other techniques). One of IETF working group's aims is to adapt the typical MPLS architecture to the Internet specifics. As a result the so called Seamless MPLS (**Figure 7.2**) has been defined [8].MPLS as a mature and well known technology is widely deployed in today's core and aggregation/metro area networks. Many metro area networks are already based on MPLS delivering Ethernet services to residential and business customers. Until now those deployments have been usually done in different domains; e.g. core and metro area networks are handled as separate MPLS domains. Seamless MPLS extends the core domain and integrates aggregation and access domains into a single MPLS domain ("Seamless MPLS"). This enables a very flexible

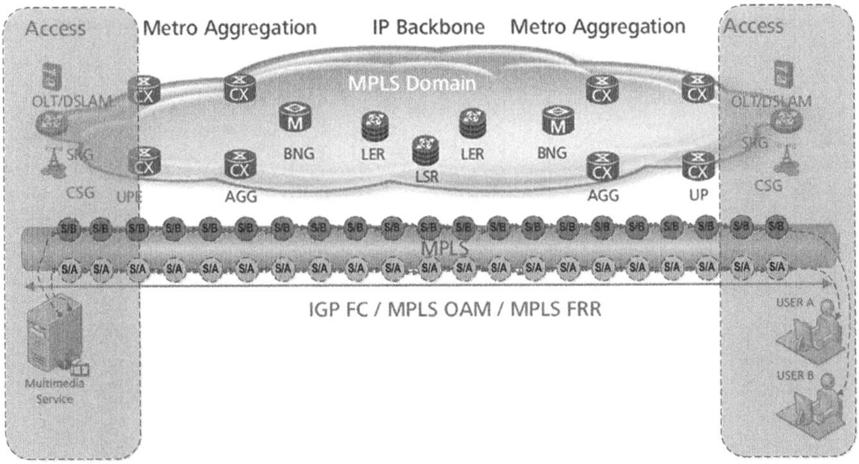

Figure 7.2 MPLS Seamless networking model

deployment of an end to end service delivery. In order to obtain a highly scalable architecture Seamless MPLS takes into account that typical access devices (DSLAMs, MSAN) are lacking some advanced MPLS features, and may have more scalability limitations. Hence access devices are kept as simple as possible [8].

Seamless MPLS describes an architecture by deploying existing protocols like Border Gateway Protocol (BGP), Label Distribution Protocol (LDP) and Intermediate System to Intermediate System (IS-IS). Multiple options are possible and IETF aims at defining a single architecture for the main function in order to ease implementation prioritization and deployments in multi vendor networks. Yet the architecture should be flexible enough to allow some level of personalization, depending on use cases, existing deployed base and requirements. For the purpose of transition to Seamless MPLS in aggregation of user traffic streams, the TR 101 specification is applied in case of a physical connection to AN on a standard telephone line (High Bit-Rate DSL - 2nd generation - HDSL2 and Single-Pair High-Speed Digital Subscriber Line – SHDSL Lines), as shown in **Figure 7.3** [9]. The TR 101 in the DSL forum draft specifies the DSL aggregation model in Ethernet mode. The model defines the V interface between the ÀN and Ethernet convergence network/node to identify different DSL ports for services and user access through the 2-layer TAG of 802.1ad (QinQ). The mainstream networking technologies of the metro convergence network include Ethernet enhanced technology (QinQ, PBB), MPLS bearer, and L3

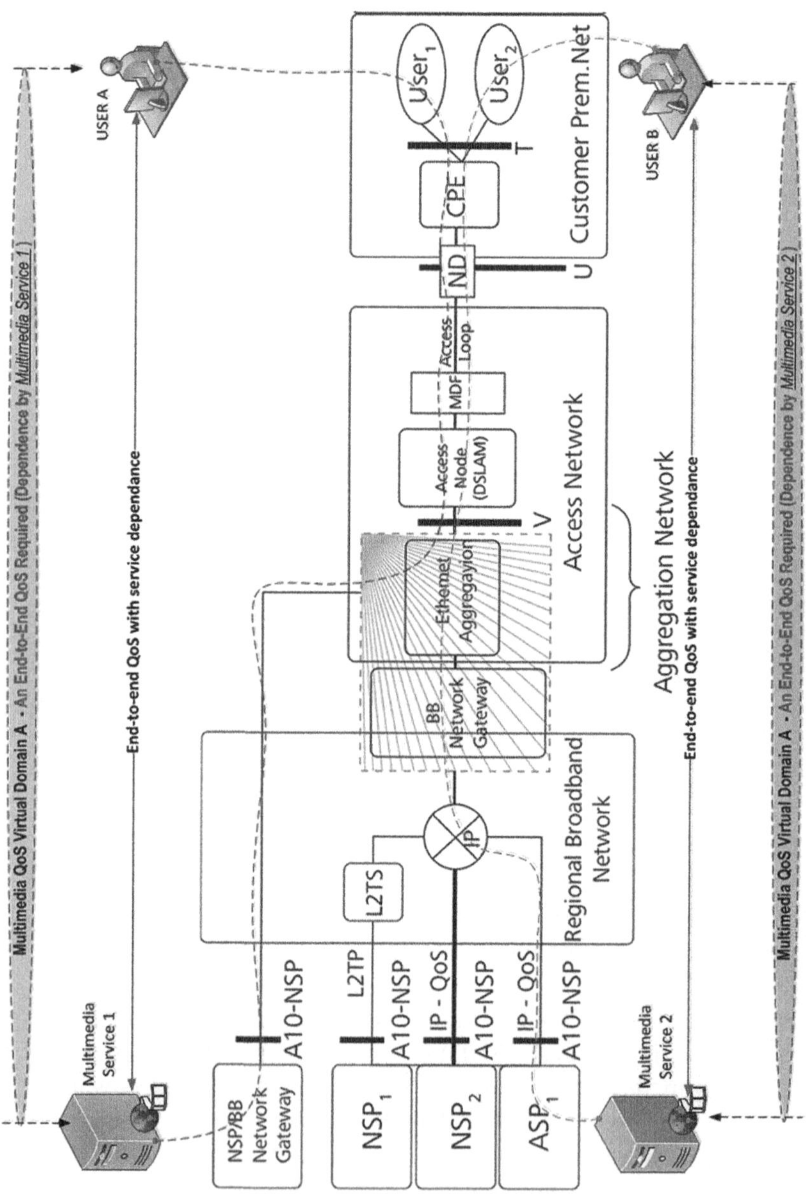

Figure 7.3 TR101 Ethernet DSL aggregation model

Hybrid. The Ethernet enhanced modes, such as QinQ and PBB, effectively improve network reliability and service flexibility. The MPLS bearer mode is one of the mainstream bearer technologies of the Ethernet convergence network, as it facilitates VLAN scalability and reliability. In L3 Hybrid mode, services are classified into edge processing services and transparently transmitted services, according to service features. For the edge processing services, the IP edge is located in the edge convergence node. Transparently transmitted services are sent to the specified POP point through the MPLS tunnel [10].

Migration from TR 101 to Seamless MPLS solves two basic problems which appear when aggregating user traffic:

- Substantially more complex mechanism for a new user session aggregation with TR101 compared to that with Seamless MPLS. From **Figure 7.4** it could be noted that the addition of a new user connection implies reconfiguration of a series of devices along the data "route", a profile change of each device, application of the new configuration, etc. These, in fact, are activities that at high dynamics of the requests consume substantial computational and communication resources.
- The lack of flexibility of TR101 when managing Access Network resources compared to Seamless MPLS. A comparison of service configuration points of Access Network resources with the TR101 architectural model and with the Seamless MPLS model is presented in **Figure 7.5**.

With Seamless MPLS a new session request or a change of parameters of an active session leads to the reconfiguration of the Customer Premises Equipment – (CPE) in the Access Nodes only, while with TR101 reconfiguration of resources must be performed in the Access Network and in the Core Network entry points as well.

Seamless MPLS development aims at provision of a possibility for effective dynamic allocation and management of key resources on the Internet at the Access Network level while achieving a high transparency level of users' connections and sessions applied in respect of architecture and Inter-Access Networks (Core Network) management.

When presenting the advantages of Seamless MPLS as the architecture of the present day it is important to estimate whether this architecture is flexible and effective enough to ensure supporting users sessions that are highly sensitive to sustainable QoS, i.e. connections and applications for multimedia information distribution and exchange over the Internet [10].

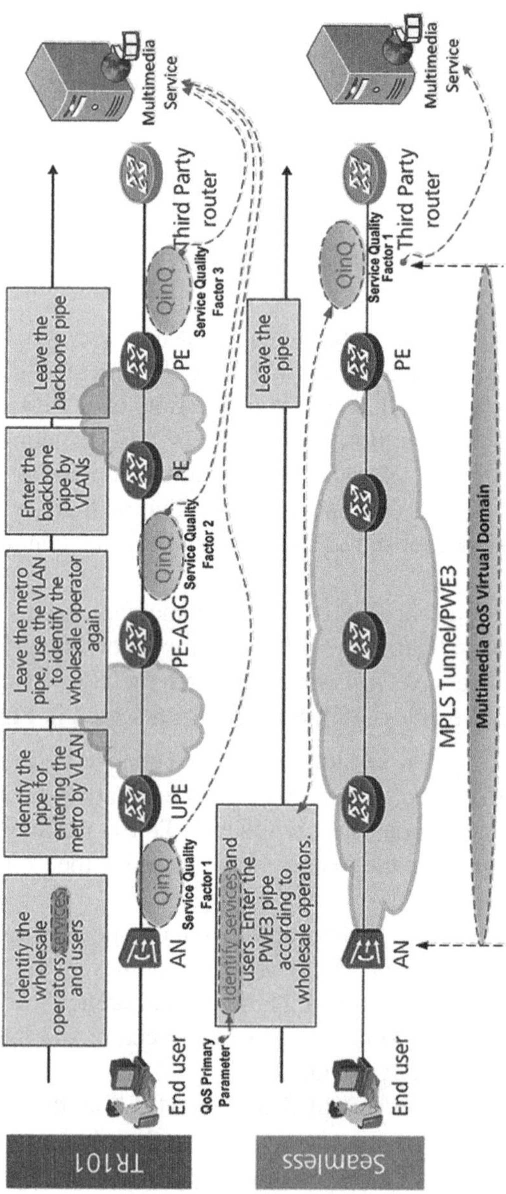

Figure 7.4 Service wholesale deployment comparison between the seamless and TR101

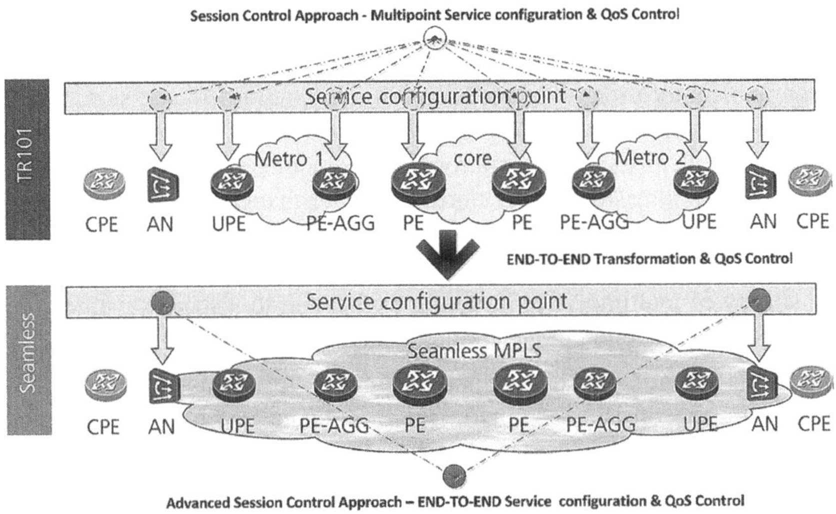

Figure 7.5 Comparison of deployment of inter-metro private line services between the Seamless and TR101

The substantial difference between multimedia and traditional data exchange sessions lies in the necessity of providing multimedia connections with [10]:

- Maximum permissible delay of multimedia packets from source IP address to target IP address.
- Minimum number of packets per unit time which must be delivered to from source IP address to target IP address which is a function of the compression method used.
- Maximum interval of time remoteness when receiving two consecutive packets which is a function of the compression method used.

These technical requirements have a substantial effect on the quality of the multimedia impact. If images are reproduced in bits and pieces and are partially non-visualized, if video and sound are not synchronized, if the whole picture "freezes" for a moment and then continues with a visible lack of part of the movie and/or video clip, if visualization preparation takes longer than the video itself, then no effective multimedia impact can be expected under these conditions.

A model for description of the Internet-based multimedia session request should be introduced for the purposes of tackling the problem of resources association to a specific multimedia connection. From implementation point of view every multimedia session can be represented as a communication service request through:

- Geographical location of the participants in the multimedia session;
- Type of media that are exchanged during the session;
- Starting point and expected duration of the session;
- Type of compression used during the exchange of multimedia information
- Connection type – unicast, multipoint unicast, multicast, broadcast;
- Sensitivity of multimedia information exchanged to a specific value of permissible % of errors in the communication channel.

Multimedia traffic modeling and provision of resources required for maintaining multimedia connections in Internet media is a problem which is tackled for over ten years, certain or other parameters of the listed above being prioritized for their importance in a number of publications. It is however important to note, that no matter how (from a technical point of view) will a multimedia session be parameterized, from the end users' viewpoint these are the sessions that are easiest and most natural for them to evaluate for their quality. In this sense, some other, user oriented characteristics of assessing the quality and/or impact of the Internet-based multimedia connections can also be specified, as:

- Session establishment time. In practice it is not as critical as it is with servicing of other Internet applications for real time information exchange. With multimedia connections it is permissible session establishment time to be increased by adding time for association of the necessary communication resources for the multimedia connection requested.
- A possibility for a prior limitation of user expectations concerning connection quality. In this case users assume that if they are on a "worse" connection to the Internet (irrespective of how we assess this quality parameter) then the multimedia connection quality will be lower (smaller video screen dimension, lower picture and sound quality, etc.)

Concerning users expectations and applications and with the aim of providing effective servicing of multimedia connections Internet Resources Management may be considered in two key aspects:

- **Static aspect.** It is connected to the topological planning, placement, transfer, enlargement of Internet resources (AN, NAS) capacities in view of the GIS information on the requested multimedia connection activity and profile. In Access Networks for example, which cover "high" multimedia activity territories and users, new AN and/or NAS can be added, the communication lines capacity can be increased, etc.

- **Dynamic aspect.** This relates to the virtualization of the MPLS domains and forming one or more MPLS domains of one or more MPLS Domain MMD depending on the current dynamics and intensity of the requested for servicing and currently serviced multimedia sessions.

The development of effective mechanisms concerning the static aspect of multimedia sessions Internet servicing is usually connected to the application of graph theory apparatus. One of the possible graph theory models which is suitable for the case is the so-called 'cascade arrangement of service stations' [11–12].

Active work has been carried out during the last three years in the field of the multimedia sessions Internet servicing as well. This activity is objectively driven by the Internet usage today. Internet usage in Europe and in the USA are presented by a source of Sandvine^{TM} [14].

7.3 Towards Multimedia Based Internet

Today the multimedia traffic generated as a result of real time entertainment and communications forms more than 60% of the total traffic in the network [14]. Data and forecasts for the USA for 2012-2018 further develop the tendency of multimedia traffic share continuous increase in comparison to the rest of data types that are shared on the Internet. This stimulates the investigations aiming at transformation of the traditional Internet architecture towards a multimedia oriented Internet architecture. **Figure 7.6** gives an idea of such a model which is represented in the overview of Kumar, Turuk [10].

The architecture in **Figure 7.6** summarizes the current status of the investigations on this problem. Key role in these investigations plays the assumption that it is possible to provide the necessary QoS for multimedia applications on the basis of protocol specifications and algorithms only. This assumption is important in view of the provision of the transition from **Internet based multimedia to multimedia based Internet**. This 'play of words' is based not only on the "Sandrive" statistical data that speak for themselves, but on the reality that we face every day as well. For example, the world game industry leader – "Blizzard" owes its success not only to its team that developed World of Warcraft^{TM} but also to the Internet which is the only communication media having the potential of providing a medium for over 4 million active on-line multimedia users. Blizzard has created its own subsystem or subnetworks

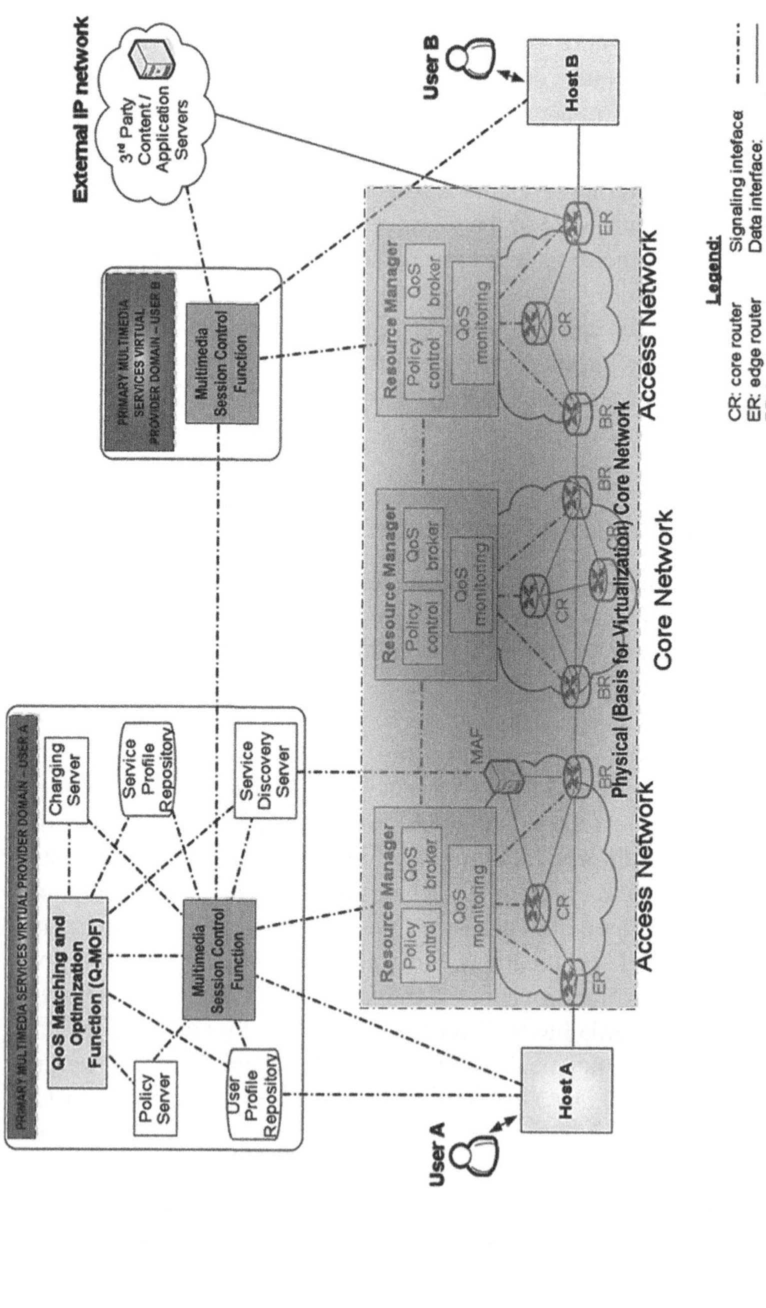

Figure 7.6 An end-to-end QoS provisioning architecture

on the Internet for this purpose (**Figure 7.7**) [15], which consists of a system of more than 245 servers of various types, as for example the so called „Low population Server", which services a geographic zone with a potential of up to 5000 real time players or "High population Server", which provides the possibility of 25, 000 gamers to play simultaneously.

It becomes clear from **Figure 7.7** that each user may choose a Game Server before the start of and/or during his multimedia session on the basis of the statistics on the current load and connectivity tests, which are embedded in the session establishment and management interface. If Word of Warcraft is a MMRPG (Massively Multiplayer Online Role-Playing Game) and practice has made Blizzard Entertainment to develop its own virtualized architecture based on the Internet resources, then and judging from Sandrive data, the modern Internet is turning into a MMUORPM – Massively Multi-User Online Role Playing Multimedia.

Figure 7.8 proposes a possible transition architecture to Internet of the MMUORPM type. The approach to the physical resources is similar to that chosen by Blizzard Entertainment, the so called virtual multimedia domains are defined within the physical Seamless MPLS architecture.

The transition model as shown in **Figure 7.8** on one hand reflects the geographic activity in respect of multimedia connection requests by domains, called Multimedia Domains, and on the other, the real opportunity for each user to request a specific type of connectivity to one and/or more NAS, called MMDC-Multimedia Domain Controller.

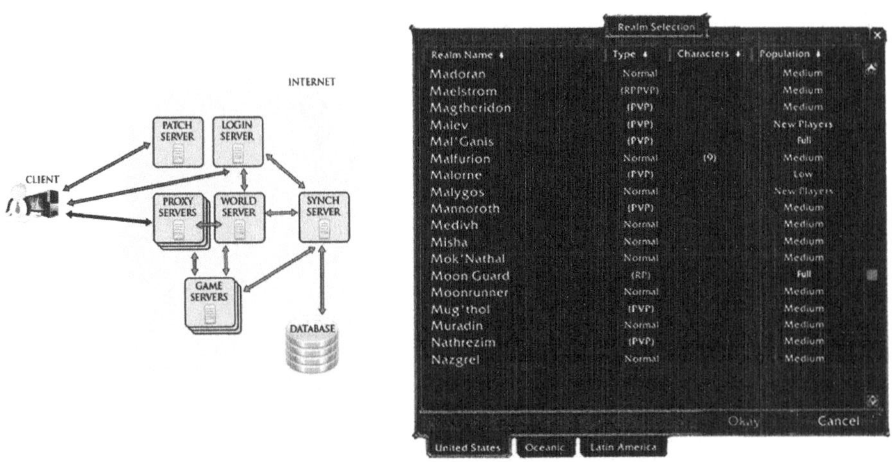

Figure 7.7 Virtual WoW network of 245 servers based on Internet resources

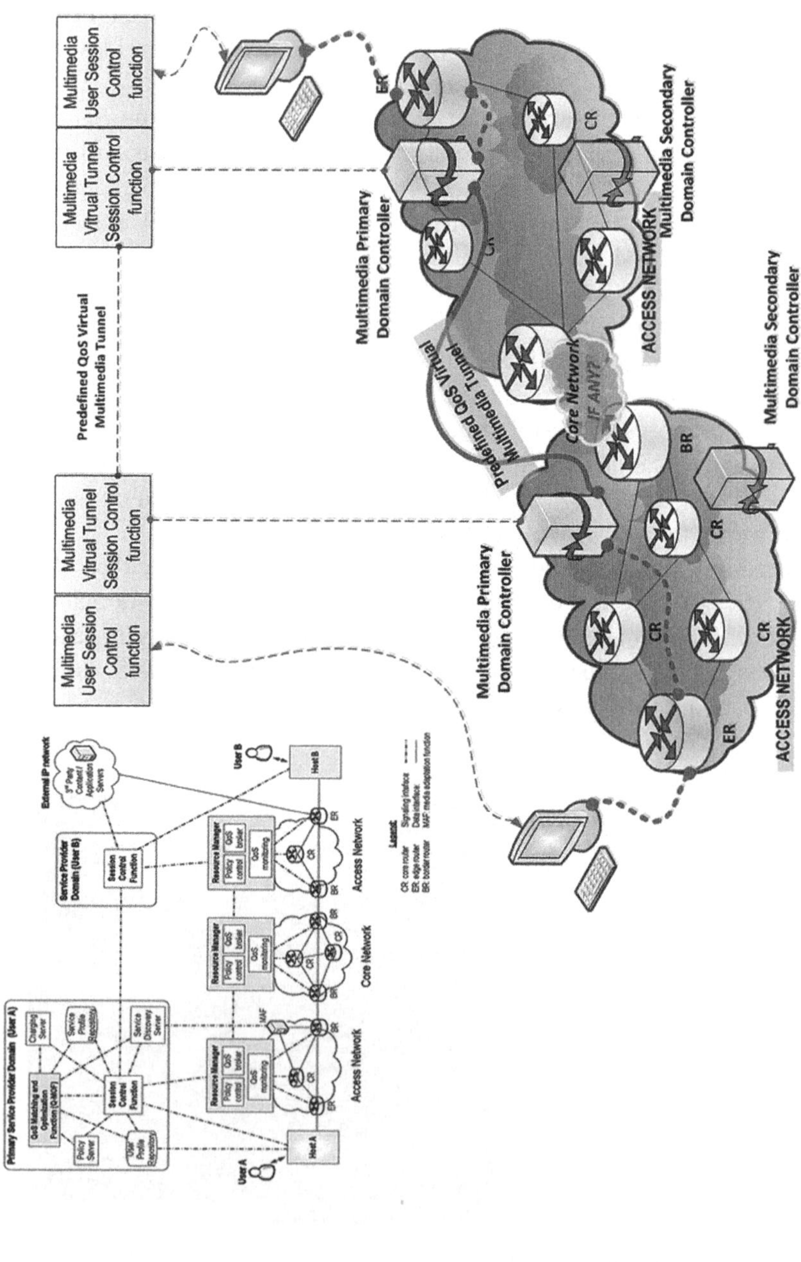

Figure 7.8 Transition to a MMUORPM Internet architecture.

The MMUORPM architectural model is a consequence of the approach of simultaneous application of the layered virtualization of physical resources on the Internet and the GIS definition of zones of "responsibility" or "activity" of servicing computers for each semantically oriented layer of services.

Application of the MMUORPM architectural model from **Figure 7.8** in fact does not impose substantial changes to the current condition of the Internet physical architecture. Additional "transparency" concerning the Core Network resources is provided by means of virtualization of the connectivity between the separate MMPDC (Multimedia Primary Domain Controller), e.g. the architectural element Core Network is missing in the virtualized application architecture of the MMUORP – Internet.

The following practical problems must be solved for the creation or generation of the virtualized application architecture of MMUORPM Internet.

Based on the statistical analysis of multimedia service applications the following must be defined:

- Geographical location (connection point in the Access Network) and MMPDC and SMPDC servicing zones (in metrical DVA remoteness from users). By 'servicing zones' we understand a number of geographically concentrated users, who systematically and with a priority make applications for multimedia sessions. When defining the range of the zones the following classification principles apply:
- Maximum remoteness between user CPE and MMPDC, measured in DVA metrics, applied for packets routing in the Access Network, in which the servicing zone is situated;
- Multimedia service applications density measured as the average value of an MMS as the relation of the total number of multimedia sessions/to total number of service applications (irrespective of service type and/or application) in the zone. Depending on the applications density, a classification of zones must be introduced, as for example:
    - Low density of multimedia service applications zone – under 30%
    - Medium density of multimedia service applications zone – from 30 to 50%
    - High density of multimedia service applications zone – from 50% to 75%
    - High intensity of multimedia service applications zone – over 75%

Depending on the zone type of the MMPDC (MMSDC), a configuration choice is made in view of the computational architecture and communication interfaces productivity.

- Mechanism for the provision of dynamic management of Predefined QoS Virtual Multimedia Tunnel (PQoS VMT) traffic characteristics (**Figure 7.5**);
- Algorithms for servicing the multimedia connections by means of the two managing functions:
    - Multimedia User Session Control function – at user-user level;
    - Multimedia Virtual Tunnel Session Control function – at „MMPDC-MMPDC" level;

What are the potential advantages of the MMUORPM Internet architecture from **Figure 7.8**?

The known models, which are used for allocation and management of resources on the Internet in connection with multimedia sessions for exchange of information (InSrerv, DiffServ and Native IP MPLS), are based on the expectation that protocol logics support in the Access Networks is provided transparently and continuously for every edge router (ER), core router (CR) and border router (BR). The realities of the constantly developing Internet, however, show that the existence of just a single ER, CR or BR which does not support the MPLS logics on the "route" of a connection is enough for the expected result not to be achieved. Due to this reason, the establishment of the multimedia session in accordance with the architecture in **Figure 7.8** is accomplished in two stages:

- **Stage one**: provision of necessary QoS and dynamic control of Predefined QoS Virtual Multimedia Tunnel (PQoS VMT) traffic characteristics on the route of the multimedia connection between subscribers A and B, applying a MPLS Tunnel (see **Figure 7.4** and **Figure 7.5**), illustrated in **Figure 7.9**;
- **Stage two**: establishment of a symmetric connections couple „End User to MMPDC", as illustrated in **Figure 7.10**.

What is the methodology for the establishment and control of the MMUORPM Internet Architecture shown in **Figure 7.8**? From a technological viewpoint, resources virtualization in the Access Networks model based on Seamless MPLS provides the solution to the problem of establishment and control of the key MMUORPM resource: Predefined QoS Virtual Multimedia Tunnel. In this sense the establishment of the MMUORPM architecture consists of:

- **Phase 1**. Definition of the necessary number and place of MMPDC (NAS) in the Access Networks

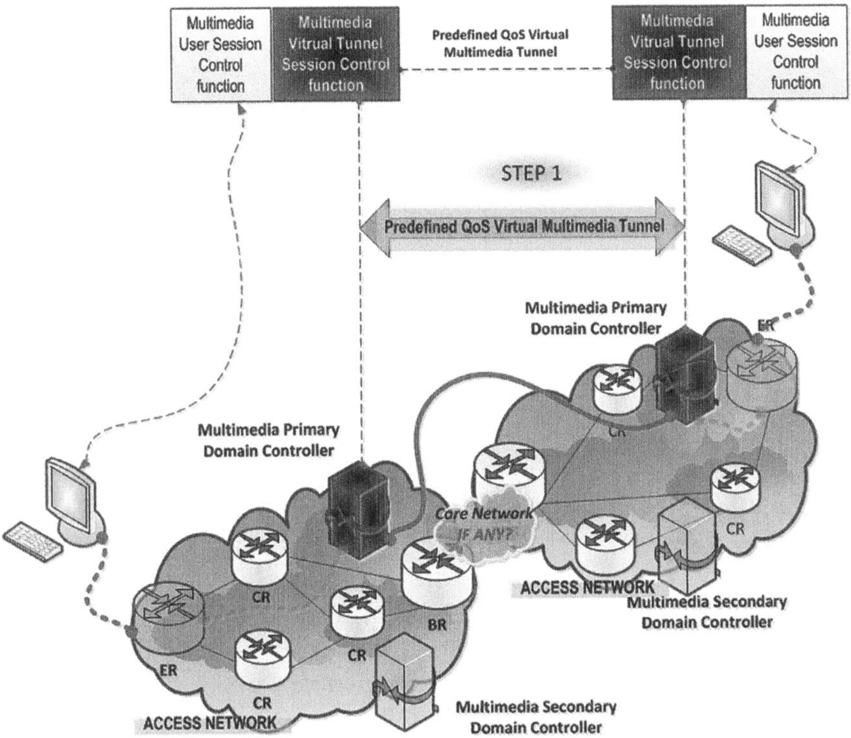

Figure 7.9 Predefined QoS Virtual Multimedia Tunnel traffic characteristics control.

- **Phase 2**. Establishment, dynamic reconfiguration and support of the necessary number of Seamless MPL Tunnels (MPLS Tunnel/PWE3 Pipes, according to **Figure 7.4** and **Figure 7.5**), on which the virtual connections are to be built and function between MMDC (defined in the model in **Figure 7.8**) as a **Predefined QoS Virtual Multimedia Tunnel** illustrated in **Figure 7.11**.
- **Phase 3**. Servicing of user requests outside the **Predefined QoS Virtual Multimedia Tunnel** – according to the algorithm presented in **Figure 7.12**
- **Phase 4**. Servicing of user requests **End-To-End.**

What problems are being solved by introducing such an MMUORPM Internet architecture?

The application of the universal model of Internet traffic streams prioritization on the basis of Seamless MPLS does not solve the key problem

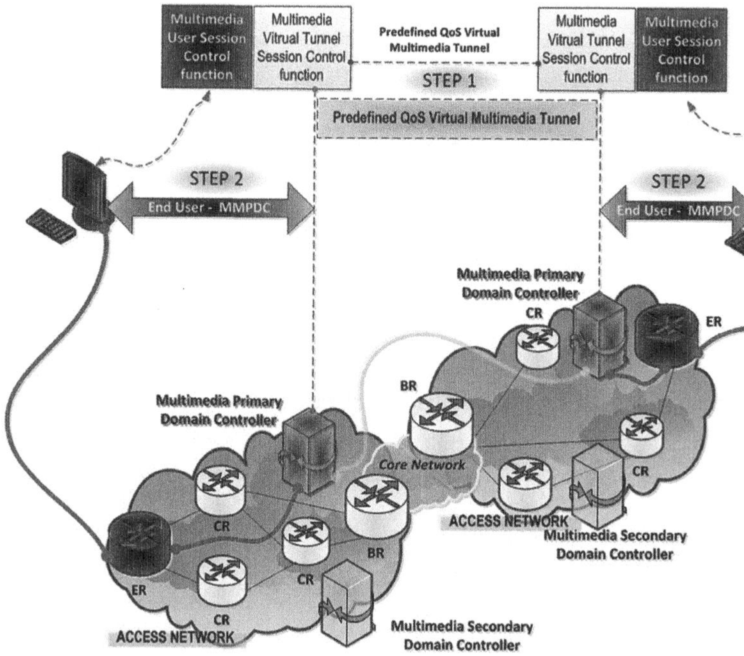

Figure 7.10 Establishment of a symmetric connections couple "End User to MMPDC".

of every management system – i.e. the level of entropy (uncertainty and disorder of the system) remains due to the fact that the management model in Seamless MPLS is of the reactive type, i.e. ''service request – adequate allocation and management of system resources based on the request". The chance of reaching the situation where the requested QoS cannot be provided is very high with this model.

The goal of the application of virtualized MMUORPM Internet architecture is a transition to proactive resources allocation and management. According to phase 1 and phase 2 of the MMUORPM architecture establishment, the topology will be formed and resources will be reserved not based on the specific multimedia servicing requests but by a statistical evaluation of the requests intensity and density for a long enough historical time interval, generated by the geographic service zones defined above. On the basis of a statistical analysis and the prediction of the multimedia users' behavior (set in phases 1 and 2) a virtual architecture is generated, which for a given QoS and minimum resources allocated, should provide service of the requests generated by each of the zones (multimedia domains) of multimedia sessions

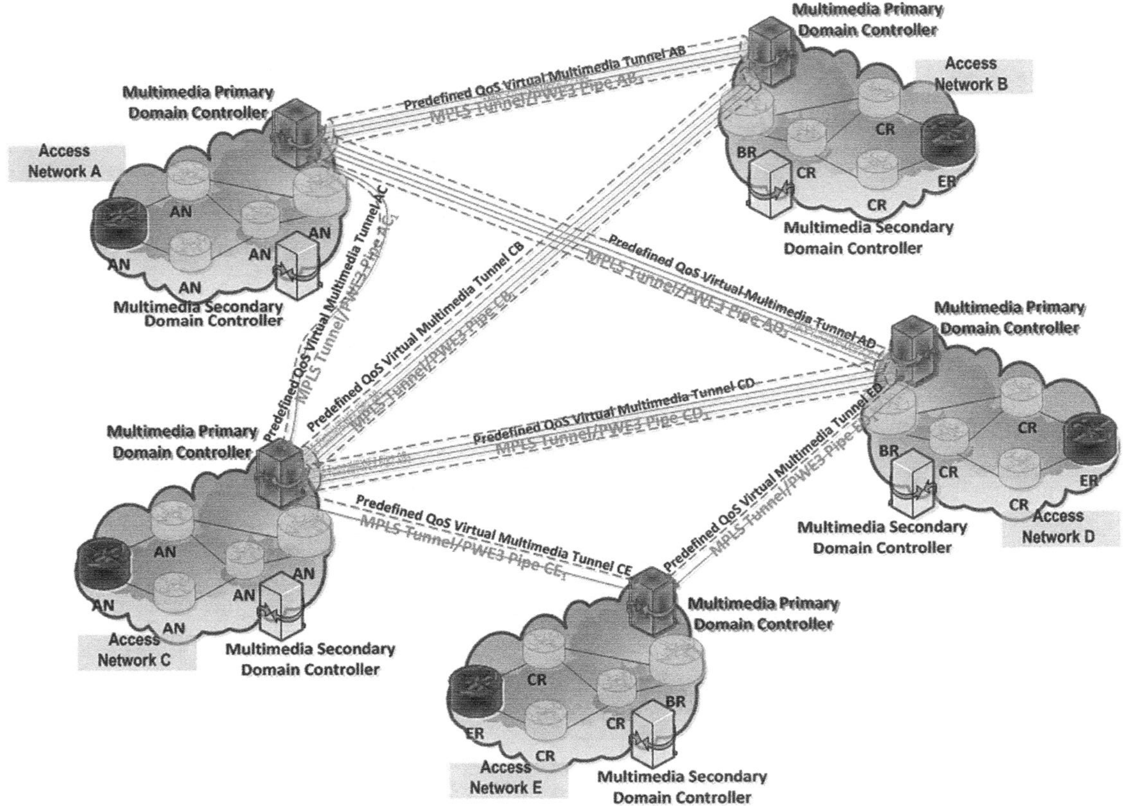

Figure 7.11 Deployment of MMPDC (NAS) and establishment of MMPDC 2 MMPDC connections.

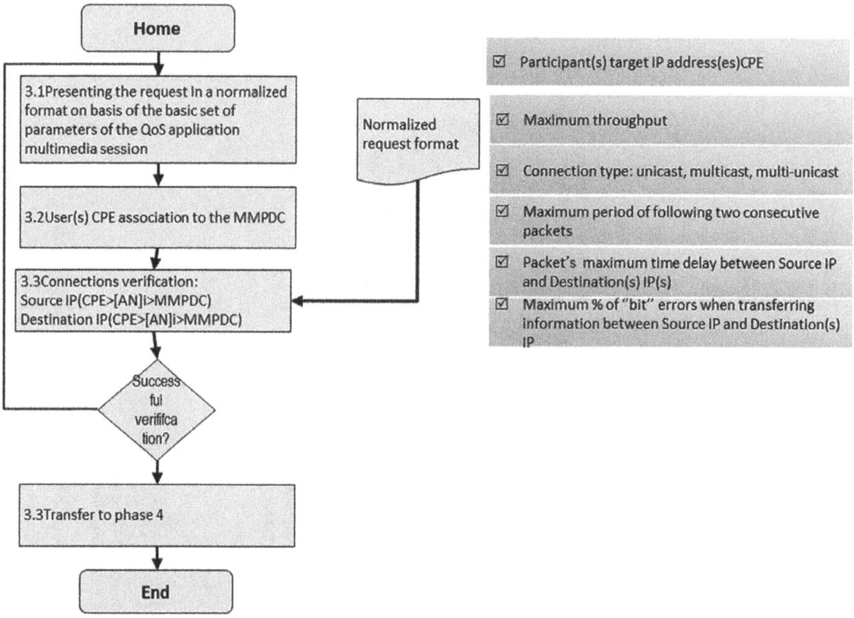

Figure 7.12 Algorithm for servicing user requests outside the **Predefined QoS Virtual Multimedia Tunnel**, according to **Figure 7.10**.

concentration. An additional resource - data base for MMUORPM architecture management (DB(MMUORPM)) should be introduced (**Figure 7.13**). This data base DB(MMUORPM) will gather and contain information concerning:

- Limits of each MMPDC's service zone, i.e. what part (Multimedia Access List) of the physical Access Network is serviced by the specific MMPDC;
- Zone type depending on multimedia services requests' density;
- Intensity and service requests in the zone for the last three successive slots (Multimedia Service Slots);
- Parameters of the **Predefined QoS Virtual Multimedia Tunnels** which connect MMPDC to the MMUORPM topology for the last three successive slots.

DB (MMUORPM) will be a distributed relation data base, a full copy of the data base being supported only in Root MMPDC (**Figure 7.12**) and in every MMPDS will be replicated only the relations connected to the specific MMPDC. In this way, even if the connection to the Root MMPDC is lost for

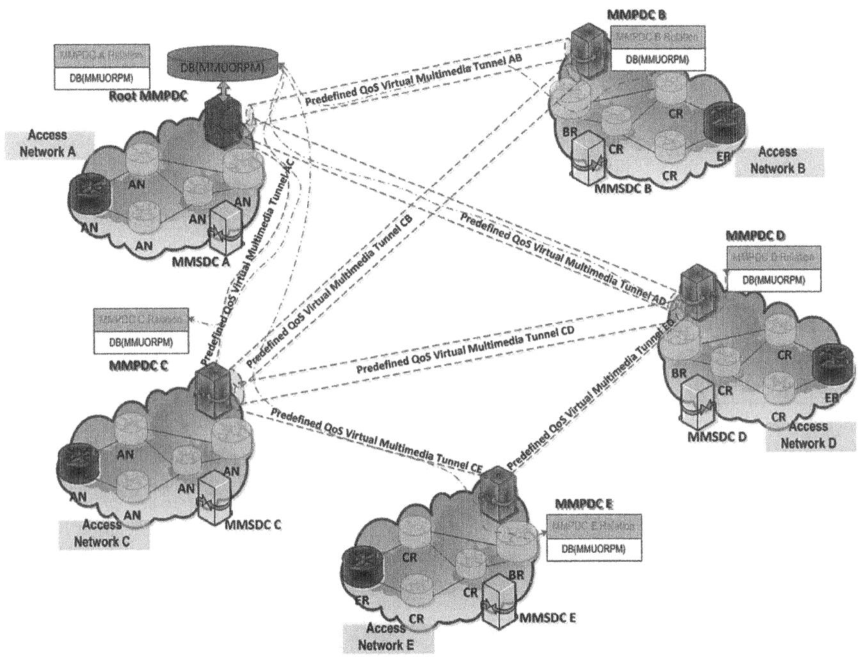

Figure 7.13 Model of data for MMUORPM virtual architecture management.

a certain moment, MMUORPM Internet architecture backbone will continue functioning.

Information from the DB (MMUORPM) will be used for the management of the MMUORPM architecture, managing the impact for the next slot (Multimedia Service Slot – MSS) being formed as:

- Adding/removing (activating/deactivating) MMPDC;
- Associating user CPE to MMPDS for servicing within the next slot;
- Establishing new and change of parameters of existing **Predefined QoS Virtual Multimedia Tunnels**.

MMUORPM Internet architecture application will substantially reduce the level of entropy and contingency in multimedia traffic servicing. MMUORPM backbone structure – the set MMPDC and **Predefined QoS Virtual Multimedia Tunnels** are quasi-stationary in view of the physical resources consumption on the Internet. In this way, the occupation of the physical resources on the Internet is "isolated" from the multimedia users' current activity. This activity affects the local connectivity – user CPE, ANs en route to servicing MMPDC.

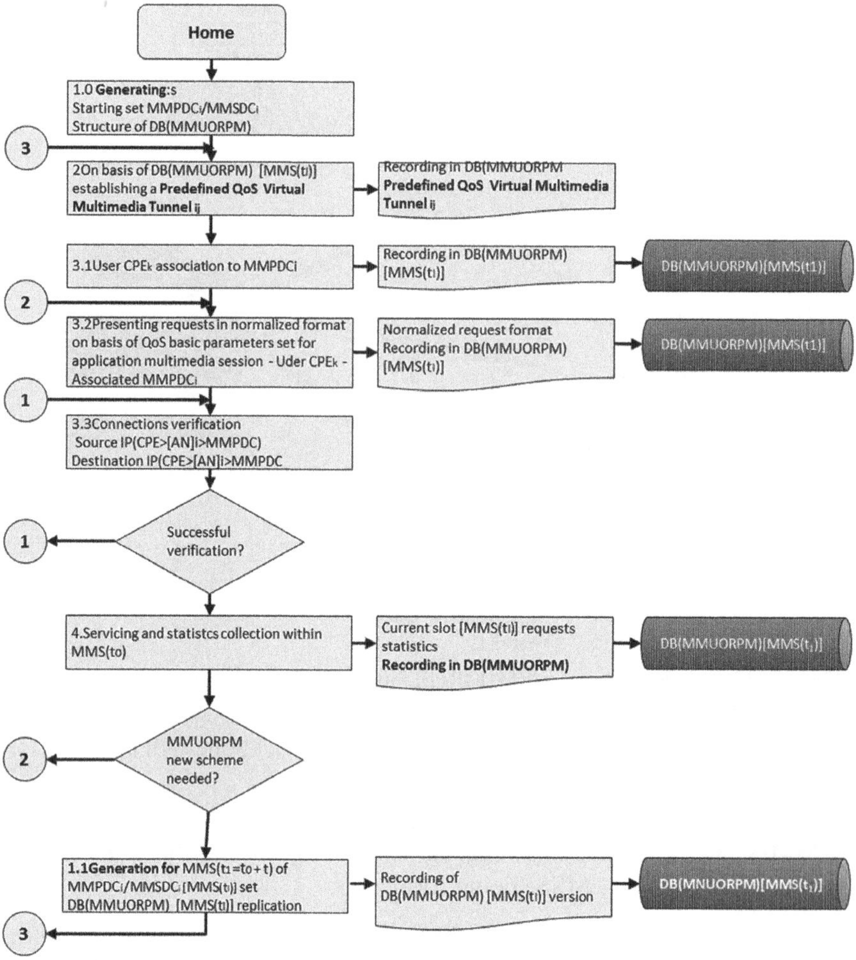

Figure 7.14 MMUORPM virtual architecture management algorithm.

The effect of isolation of the multimedia users' current activity in respect to the occupation of the physical resources on the Internet has also an additional advantage. It provides the possibility to limit the impact of the multimedia traffic on the provision of QoS for other critical services, which are increasingly based on Internet connectivity, such as Point-to-Point tunnels for real-time data collecting systems, sensor networks, critical information infrastructure applications, etc. The application of MMUORPM Internet architecture confines occupation of physical resources and their management to the Multimedia Service Slot – MSS level. The longer the duration of the MSS is, the smaller the "expenses" are for service traffic and computational resources for MMUORPM virtual architecture and for the management of the physical resources on the Internet for provision of the MMUORPM virtual architecture itself.

7.4 Conclusions

With the growth of service and application demands pushed by the convergence of technologies and enabling new forms of social and business interactions, the Internet faces the important challenge to support sustainably the increased information flows also enabled by the higher capacity and lower delays at wireless connectivity level that are provided by next generation mobile communication systems. This Chapter proposed a transition Internet architecture that can substantially reduce the level of entropy and contingency in multimedia traffic servicing, while enabling to carry increasing traffic and manage more dynamic multimedia demands. The methodology for the establishment and control of the proposed architecture was also described. A key element enabling efficiency of resource allocation, and QoS provision, in particular for critical services hugely dependent on proper Internet connectivity (e.g., processing of sensor data) is the Predefined QoS Virtual Multimedia Tunnel. This element is also the key enabler to proactive resources allocation and management as it allows for interoperable decentralized Internet service-oriented management and control.

Cost-and energy-efficiency are other critical challenges for the Internet-based communication infrastructure of today and will even be more prominent in the future. The proposed management algorithm and concept allows to reduce the cost of traffic servicing and computational resources.

The future work envisions the concept to be tested through the development and approbation of a fully virtualized platform, a server part with an integrated agent for dynamic multimedia management and a client (plug-in) for the most commonly used browsers. Based on this platform an performance analysis will be done in relation to the possibilities for application of the proposed approach for different profiles of the requests for multimedia service, as well as other types of real-timed managed services realized on internet based connections.

References

[1] Access Node Control Protocol (ANCP) MIB module for Access Nodes, Standards Track Active Internet –Draft, IETF, NETWORK WORKING GROUP, Issue Date June 24, 2013 : http://datatracker.ietf.org/doc/draft-ietf-ancp-mib-an/.

[2] Access Node Control Protocol. Cisco Reference Guide, June 25, 2009 Cisco Systems, Inc. http://www.cisco.com/en/US/docs/ios/ios_xe/ancp /configuration/guide/ancp_xe.pdf.

[3] Access Node Control Protocol for IGMP. Juniper Technical Doc.Multicast Routing Configuration Guide. Release 14.3.x, 2013 Juniper Networks, Inc., http://www.juniper.net/techpubs/en_US/junose14.3/ information-products/topic-collections/swconfig-multicast-routing/index.html?topic-70109.htm.

[4] Seamless MPLS Networking. Technical White Paper, 2010 Huawei Technologies Co., Ltd. http://www.huawei.com/en/static/hw-076762.pdf.

[5] McCloghrie, Rose M., Management Information Base for Network Management of TCP/IP-based Internets. Performance Systems International Editors, March 1999. http://www.ietf.org/rfc/rfc1213.txt.

[6] Network layer protocols, RFC Source Book, 2012 Network Sorcery, Inc. http://www.networksorcery.com/enp/topic/ipsuite.htm.

[7] Multiprotocol Label Switching. Standards Release Notes, IETF, NETWORK WORKING GROUP, Version 4.83, released 06 Nov 2013. http://datatracker.ietf.org/wg/mpls/charter/.

[8] Leymann .N, Konstantynowicz, M, Steinberg D., Seamless MPLS Architecture. Internet-Draft Document, MPLS Working Group, March 12, 2012, http://tools.ietf.org/pdf/draft-ietf-mpls-seamless-mpls-01.pdf.

[9] Cohen A. Allan D. Migration to Ethernet-Based DSL Aggregation, Technical Report, DSL Forum TR-101 Architecture and Transport Working Group,April 2006http://datatracker.ietf.org/documents/LIAISON /file468.pdf.

[10] Rajeev Kumar , J. S. Rao , A. K. Turuk , S. Chattopadhyay , G. K. Rao, A Protocol to Support QoS for Multimedia Traffic over Internet with Transcoding. Trusted Internet Workshop 2002. December 18, 2002, Taj Residency. http://citeseerx.ist.psu.edu/viewdoc/download?doi=10.1.1.93.3631&rep=rep1&type=pdf.

[11] J. Reese. Methods for Solving the p-Median Problem: An Annotated Bibliography. Research Report, Department of Mathematics, Trinity

University, August 2005. http://ramanujan.math.trinity.edu/tumath / research /reports/report96.pdf.

[12] Yu An, Bo Zeng, Yu Zhang and Long Zhao, Reliable p-median facility location problem: two-stage robust models and algorithms, Research Report , University of South Florida, Tampa, FL 33620, December, 2012. http://www.optimization-online.org/DB_FILE/2012/12/3712.pdf.

[13] A Journey Into MMO Server Architecture Posted by FaceOfMankind Thursday May 30 2013. http://www.mmorpg.com/blogs/FaceOfMankind /052013/25185_A-Journey-Into-MMO-Server-Architecture.

[14] Sandvine global internet phenomena report; 1H 2013. https: //www.sandvine.com/downloads/general/global-internet-phenomena /2013/sandvine-global-internet-phenomena-report-1h-2013.pdf.

[15] WOW Server List - US & EU Realm Status, Information Notice, 2011 wowrealmstatus. http://www.wowrealmstatus.net.

Index

About the Editors

Vladimir Poulkov is currently Professor and Dean of the Faculty of Telecommunications at the Technical University of Sofia, Bulgaria. He has received his MSc and PhD degrees at the Technical University of Sofia. He has more than 30 years of teaching and research experience in the field of telecommunications. Major fields of scientific interest are in the fields of Information Transmission Theory, Modulation and Coding. His has expertise related to interference suppression, power control and resource management for next generation telecommunications networks. He is a Senior Member of the Institute of Electrical and Electronic Engineers (IEEE), thematic area leader in "Embedded and Resource Optimal ICT" at the Center for TeleInfrastruktur (CTIF), chairman of Bulgarian Cluster of Telecommunications.

Ramjee Prasad is currently the Director of the Center for TeleInfrastruktur (CTIF) at Aalborg University, Denmark and Professor, Wireless Information Multimedia Communication Chair. Ramjee Prasad is the Founding Chairman of the Global ICT Standardisation Forum for India (GISFI: www.gisfi.org) established in 2009. GISFI has the purpose of increasing of the collaboration between European, Indian, Japanese, North-American and other worldwide standardization activities in the area of Information and Communication Technology (ICT) and related application areas. He was the Founding Chairman of the HERMES Partnership - a network of leading independent European research centres established in 1997, of which he is now the Honorary Chair. He is the founding editor-in-chief of the Springer International Journal on Wireless Personal Communications. He is a member of the editorial board of other renowned international journals including those of River Publishers. Ramjee Prasad is a member of the Steering, Advisory, and Technical Program committees of many renowned annual international conferences including Wireless Personal Multimedia Communications Symposium (WPMC) and Wireless VITAE. He is a Fellow of the Institute of Electrical and Electronic Engineers (IEEE), USA, the Institution of Electronics and

Telecommunications Engineers (IETE), India, the Institution of Engineering and Technology (IET), UK, and a member of the Netherlands Electronics and Radio Society (NERG), and the Danish Engineering Society (IDA). He is a Knight ("Ridder") of the Order of Dannebrog (2010), a distinguished award by the Queen of Denmark.

About the Authors

Ivaylo Atanasov received his Master degree in electronics from Technical University of Sofia, and PhD degree in communication networks. His current position is Professor at Faculty of Telecommunications, Technical University of Sofia. His main research focus is on Internet protocols, multimedia communications and resource management. He has experiences in development of open service platforms for next generation networks.

Oleg Asenov, has received his MSc degree at the Technical University of Gabrovo and PhD degree at the Technical University of Sofia. He has more than 20 years of teaching and research experience in the field of Telecommunications. The major fields of scientific interest are modeling, simulation and design of computer networks based on graph theory and applied heuristics algorithms. Currently he is Associated Professor at the St.Cyril and St.Methodius University of Veliko Tyrnovo, Member of IEEE.

Todor Cooklev received his PhD degree from Tokyo Institute of Technology in 1995. Currently he is Director of the Wireless Technology Center at Indiana University-Purdue University Fort Wayne, Fort Wayne, Indiana, and ITT Associate Professor of Wireless Communication and Applied Research at the same institution. His research interests are in most aspects of modern wireless systems. He has given presentations, talks, and short courses worldwide. He has contributed to about 100 publications and several patents in the United States and other countries. He has participated in several standards organizations developing commercial and tactical wireless communication systems.

Anton Hristozov graduated from the Technical University in Sofia Bulgaria in 1989. His first position was a research associate at the Institute of Robotics of the Bulgarian Academy of Sciences. Later he obtained a Master's Degree in Telecommunications and Information Science from the University of Pittsburgh. During the last 25 years he has held several technical positions in the fields of Telecom, Robotics and Transportation. Anton currently lives in Pittsburgh, Pennsylvania where he works for NetApp as a firmware engineer. His expertise is in creating robust embedded systems which have different sensors and communication protocols.

Kiril Kassev holds B.Sc., M.Sc., and Ph.D. degrees from the Department of Communication Networks at the Technical University of Sofia. He is currently an assistant professor at the Department of Communication Networks. Dr. Kassev has been involved as a researcher and consultant in national and EU-funded projects. He has published a number of papers in referred journals and presented in various national and international conferences. His research interests fall in the area of performance analysis and QoS provisioning algorithms in wireless networks.

Sofoklis Kyriazakos graduated Athens College school in 1993 and obtained his Master's degree in Electrical Engineering and Telecommunications in RWTH Aachen, Germany in 1999. Then he moved to the National Technical University of Athens, where he obtained his Ph.D. in Telecommunications in 2003. He also received an MBA degree in Techno-economic systems from the same university. He has more than 90 publications in international conferences, journals, books and standardization bodies. He has been invited as reviewer, chairman, member of the committee, panelist and speaker in many conferences and has also served as TPC chair in 2 International conferences. Currently he holds the academic position of Associate Professor in the University of Aalborg. Sofoklis has managed, both as technical manager and coordinator, a large number of multi-million ICT projects, at R&D and industrial level. In 2006 Sofoklis founded an ICT startup, Converge S.A., with the PRC Group and since then he has been the Managing Director and a BoD member. Since March 2012, he also Managing Director and BoD member of Converge ICT Innovation Inc. based in Montreal, Canada, which is an affiliate company. Sofoklis is also member of the BoD of Athens Information Technology, a Center of Excellence for Research and Education.

Garik Markarian received a First Class Honours degree in radio communications in 1976, and the Ph.D. degree in communication systems in 1982, both from Odessa Technical University, Telecommunications, Odessa, Ukraine. He is a Professor at Lancaster University and holds a Chair in communication systems at the School of Computing and Communications,

Lancaster University, Lancaster, U.K. His research interests include wireless broadband communications, e-health, security, and video distribution over wireless networks. He co-authored of more than 300 publications, including 4 books, 42 patents, and great number of papers in leading journals.

Lyudmila Mihaylova is an Associate Professor (Reader in Advanced Signal Processing and Control) with the Department of Automatic Control and Systems Engineering, University of Sheffield, United Kingdom. Her interests are in the area of nonlinear filtering, sequential Monte Carlo Methods, statistical signal processing and sensor data fusion. Her work involves the development of novel techniques, e.g. for high dimensional problems (including for vehicular traffic flow estimation and for image processing), localisation and positioning in sensor networks. Dr. Mihaylova is an Associate Editor of the IEEE Transactions on Aerospace and Electronic Systems, of Elsevier Signal Processing Journal and the Editor-in-Chief of the Open Transportation Journal and. Dr. Mihaylova is a senior member of the IEEE, Signal Processing Society and a member of the International Society of Information Fusion (ISIF) and an ISIF board member.

She has been serving to the scientific community also as a member of the Programme/ Organising Committee of international conferences and symposia, including the International Conferences on Information Fusion, the American Control Conferences, EUSIPCO, conferences on Intelligent Transportation Systems and the German workshops on Multiple Sensor Data Fusion. Dr Mihaylova has given a number of invited talks, e.g. the keynote speech for the 5th IET International Conference on Wireless, Mobile and Multimedia Networks (2013), Beijing, China, tutorials including for the EU Marie Curie ITN (2010, Sweden, 2012, Germany) and COST-NEARCTIS workshop (2010, Switzerland). Her research is funded by sponsors such as EPSRC, EU, MOD and industry.

Yakim Mihov received his B.Sc. and M.Sc. degrees in Telecommunications from the Technical University of Sofia, Bulgaria, in 2008 and 2010, respectively. He graduated with a Ph.D. degree in Telecommunications from the same university in 2013. He has participated in research projects. Dr. Mihov has served as a technical program committee (TPC) member of international conferences and as a reviewer for international journals (such as Journal of Information Processing Systems, Computer Networks (Elsevier), IEEE

Wireless Communications Magazine, IEEE Communications Letters, IEEE Systems Journal, etc.). His research interests are in the field of wireless communications systems and include (but are not limited to) cross-layer design, power saving mechanisms, optimal resource allocation, and quality of service (QoS) provisioning in broadband wireless access networks. His current research work focuses on QoS provisioning in cognitive radio networks.

Albena Mihovska has a PhD degree from Aalborg University, Aalborg, Denmark where she is currently an Associate Professor and Head of Standardisation at CTIF. She has a strong radio communications background and is currently involved in research related to integration of radio communications into the cloud and Internet of Things scenarios for optimizing and supporting reliable and high performance intensive data rate communications. She has been active within ITU-T GSI-IoT standardization.

Evelina Pencheva received her Master degree in mathematics from University of Sofia, Bulgaria, and PhD degree in communication networks from Technical University of Sofia. Her current position is Professor at Faculty of Telecommunications, Technical University of Sofia. Her interests include multimedia communications, service delivery platforms and network protocols. She has experiences in development of next generation mobile applications and middleware platforms.

Denis Rodionov received the master degree in Bauman Moscow State Technical University in computer sciences in 2003. He has software engineer experience in web technologies, storage systems and video surveillance since 2000. Research career has been started from 2010 when he started PhD in communication technologies in Lancaster University. His research interests are indoor location and positioning, location prediction and location fingerprinting technique.

John Soldatos holds a BSc. degree and a PhD degree (2000) both from the National University of Athens. Since 1996 he has had very active involvement in more than fifteen research projects in the areas of broadband networks, pervasive/cloud computing, and the internet-of-things. He is the initiator and co-founder of open source projects AspireRFID (http://wiki.aspire.ow2.org) and OpenIoT (https://github.com/OpenIotOrg/openiot). He has published more than 140 papers in international journals and conferences. Since 2003 he is with Athens Information Technology, where he is currently an Associate Professor. He has also been an Adjunct Professor at the Information Networking Institute (of the Carnegie Mellon University (2007–2010) and a Honorary Research Fellow of the School of Computing of University of Glasgow (2014–2015).

Boris Tsankov graduated with a M.Sc. degree in Communication Engineering from Technical University of Sofia, Bulgaria, and a Ph.D. degree from Moscow University of Informatics and Communications, Russia. Prof. Tsankov is an author and co-author of 10 books on Teletraffic Engineering and Fundamentals of Telecommunications. He has published more than 100 research papers in referred journals, national and international conferences, and as an invited speaker. He has served on the technical and executive committees of several conferences. Prof. Tsankov has been engaged as a consultant to a number of telecom companies and research projects' teams. He has also been involved in many national and international projects as a coordinator and experienced team leader. His research interests include teletraffic engineering and QoS provisioning in fixed, mobile, and IP networks.

Anna Zvikhachevskaya received her Ph.D. degree in Communication Systems Engineering from Lancaster University, UK in 2010, M.Sc. and B.Eng. from State Technical University (Orel, Russia) in 2007 and 2005 respectively. Since 2007 she have been involved in R&D related to the eHealth/mHeath solutions, wireless communication systems, signal processing, estimation, tracking, data fusion, algorithm engineering and data analysis. She began her career as a Research Associate in 2010 at Lancaster University, UK.

Dr. Zvikhachevskaya's PhD Research work was dedicated to the development of advanced communication networks and protocols adopted for e-Health and Telemedicine applications. The results of this research were also incorporated in the Cross-Layer Ad-hoc Network System which was developed by Lancaster University for UK MoD. During her employment as RA on EU FP7 Marie Currie Project "MC IMPULSE" she developed and optimised a novel hybrid positioning technique for wireless networks. From 2012 until summer 2013, Senior Scientist and member of the Modelling and Simulation team at the Life Scan Scotland, Johnson & Johnson's, diabetes healthcare franchise. Research work was dedicated to the development of new algorithms and inventions for the blood glucose monitoring sensors. Numerous patents and invention disclosures were generated which helped with performance evaluation and improvement of the product. Currently, she is an Algorithm Engineer and works in the R&D in Accunostics Ltd. Company.

Lightning Source UK Ltd.
Milton Keynes UK
UKOW07n2313110115

244268UK00001B/25/P

9 788793 102446